TINKERING

For Adis Hondo

TINKERING

AUSTRALIANS REINVENT DIY CULTURE

Katherine Wilson

Monash University Publishing
Matheson Library and Information Services Building
40 Exhibition Walk
Monash University
Clayton, Victoria 3800, Australia

www.publishing.monash.edu

Monash University Publishing brings to the world publications which advance the best traditions of humane and enlightened thought. Monash University Publishing titles pass through a rigorous process of independent peer review.

www.publishing.monash.edu/books/t-9781925495478.html

#tinkeringaustralia

Series: Monash Studies in Australian Society

Internal design: Les Thomas

Cover design: Mal Padgett

Cover images: *Rope Hammer*, Mark Thomson. *Wire Can*, Mark Thomson & Chris Block.

Inside cover image: Detail of Daniel Moynihan's *H W Tinker Truck*, 2010, Oil on Linen 204 cm x 275 cm. Daniel Moynihan is represented by Australian Galleries and registered with Viscopy.

National Library of Australia Cataloguing-in-Publication entry

Creator:	Wilson, Katherine, author.
Title:	Tinkering : Australians reinvent DIY culture / Katherine Wilson.
ISBN:	9781925495478 (paperback)
Subjects:	Self-service (Economics)--Australia. Do-it-yourself work Repairing--Australia. Australians--Attitudes--21st century.

Printed in Australia by Griffin Press an Accredited ISO AS/NZS 14001:2004 Environmental Management System printer.

CONTENTS

Glossary

Additive manufacturing (AM) A mechanised process of making 3D objects through layer-by-layer deposition (3D printing).

ATO Australian Taxation Office.

ABS Australian Bureau of Statistics.

Bot Shorthand for Maker Bot or other 3D printer.

Bushpunk Often associated with African low-tech utilitarian tinkering, but also an Australian variant of steampunk.

CAM Computer aided manufacturing.

Case-mod Abbreviation of case-modification: the practice of 'blinging' technologies without changing their internal workings.

Copyleft A practice and political movement for making a creative work freely available to be modified and extended.

Craftivism Popularly understood as activism using handcrafts that make political statements in public spaces, exemplified in the yarn bombing movement.

Creative industries A contested term used by sociologists and policy makers from the 1960s onward. In this book's context, used to define a sector working in cultural production.

CSIRO Commonwealth Scientific and Industrial Research Organisation.

DIY Do-it-yourself.

DIT Do-it-together.

Etsy An online trading site for handmade products.

Exhibitionary complex A contested term originating from sociologist Tony Bennett's 1995 exhibitionary complex theory, which described the system of disciplinary and power relations that emerged from the 19th-century movement from private to public display of objects and knowledges. In this book, a set of curatorial and regulatory norms and values imposed by exhibitionary institutions.

Exploratorium A science and tinkering lab based in San Francisco and online.

FabLab Membership-based community fabrication workshop.

Fixers Collective A Brooklyn-based skills-sharing project dedicated to 'creative caring towards the objects in our lives'.

GFC The global financial crisis of 2008.

Gravfest The Macclesfield Gravity Festival for downhill cart races.

Hackerspace/Repair café/Techshop Membership-based community fix and fabrication workshops.

Haptic feedback In this book the term refers to the correspondence or communication between human perception and material forces.

HP Hewlett-Packard.

IBYS Institute for Backyard Studies (South Australia).

iFixit Online free repair manual and forum.

LARP Live action role playing.

LETS Local Energy Trading Systems. A cashless system of exchange of skills, labour and produce.

Maker Faire Makers' exhibition forum, now international, founded by *Make* publisher Dale Doherty.

Maker movement A contemporary DIY culture, or imagined public, presented as grassroots and promoted by *Make* magazine and also by corporate multinationals vested in additive technologies.

MIT Massachusetts Institute of Technology.

MIY Make-it-yourself.

MONA Tasmania's Museum of Old and New Art.

P2P Peer-to-peer (sometimes person-to-person).

Recursive/recursion In this book these terms refer to the cummulative effects of repetitive recirculation of ideas and practices.

Reprap Replicating rapid prototype, an open-source project and additive manufacturing machine whose own parts and spare parts can be 3D printed by the machine itself.

Satsang A word originating in Sanscrit, meaning a companionable generation of wisdom, truth or spirituality.

Steampunk A neo-Luddite, neo-Victorian subculture with technofetish interests.

TAI The Australia Institute. An independent research body with politically progressive ideals.

TechShop America's largest fabricating franchise that gives members (customers) access to tools and workshop space.

Technocrat Contested and sometimes pejorative term invoking a political class comprising a bureaucratic, managerial, technical and industrial elite. Outside of direct quotes within this book, the term 'technocratic' is used to mean a set of managerial, regulatory, technical and bureaucratic norms that sit at odds with tinkering values.

Thingiverse A website hosting 3D printed objects and providing downloadable open-source hardware designs licensed under creative commons or general public licences.

Upcycling The transformation of waste products into usable products.

Wunderkammer A German compound word, frequently used by steampunk tinkerers, meaning a cabinet of curiosities.

Yarn bombing Textile graffiti, very often knitted or crocheted yarn fibre made over public objects.

Acknowledgements

Such an epic network of people collaborated to build this book that it seems fraudulent to claim sole authorship. I'm very privileged to have met and worked with the 32 tinkerers who welcomed me into their worlds. Many of their stories made their way into this book, and seven became its protagonists. Of these, Mark Thomson stands out for his generosity in sharing deep knowledge. In fact, this book is Mark's bumper-sticker (*I tinker therefore I am*) writ verbose.

Alongside the assembly of scholarly giants whose tools I've pinched, I owe a special debt of gratitude to Adis Hondo and Gilda Civitico, and also to the Banks and Keath families whose epic tinkering stories, tremendous as they are, didn't fully survive the final cut. My heartfelt gratitude is with Ellen Koshland, whose loving support, kindred spirit and wholesome intellect sustained this project.

Thank you to the scholarly tinkerer Ramon Lobato for your generous enthusiasm, brilliant insights and warm support during my doctoral misadventures and beyond. Thank you Julian Thomas for your humorous reproaches and critical wits. Thank you Klaus Neumann for deepening my understanding of scholarly rigour. I'm grateful for the support of the erstwhile Swinburne Institute for Social Research and its collegial community: Skye Krichauff, Ian McShane, Josephine Rasch, Anneka Lems, Steffi Scherr, Aneta Podkalicka, Michaela Callaghan, Amanda Lawrence, Ellie Rennie, Denise Merydith, Jasmina Kijevcanin, and Alexandra Heller-Nicholas.

Nathan Hollier, I'm very grateful for your wisdom, friendship and vision. And thanks to the hardworking team at Monash University Publishing, including Sarah Cannon, Les Thomas, Laura McNicol Smith, Duncan Fardon, Jo Mullins and John Mahony, who gracefully ushered this book into existence. Susan Luckman, Kirsty Robertson, Jeff Sparrow and Mark Davis, thanks for generously offering criticism. For early discussions and ideas, thank you Dan Leach, Tim Thornton, Jeff Sparrow, Liam Ward, Michael Leach, Andrew Peel, Martin Thomas, Jo Lampert, Gyorgy Scrinis, Anna Salleh and Jess McCaughey. For their various ways of helping this project, thank you Suze Houghton, Alicia Sometimes, Kat Jungnickel, Jenny Drummond, Claire Fischer, Ian Russell, Stella Glorie, Bev Hanson, John Hanson, Sharmila Nezovic, Sulja Nezovic, Nina Keath, Gary Freed, James McCaughey, Daniel Moynihan, Charlie Moynihan, Rachel Blake and Patrick Witton.

Geoff and Mary Wilson gave me love, encouragement and support from afar; Virginia Golding kept me calm; O and V tolerated a mother who wasn't always fully present; and the gadgeteer Mal Padgett cared for my family and home with incessant fun, love and heroic endurance.

Edited versions of chapter sections first appeared in *Meanjin*, *Overland*, *The Conversation* and Fairfax newspapers. A style note: in journalistic convention, protagonists' surnames are used, but in cases where their partners share their surnames, these partners are described by their first names. (So, for example, John Tucker is described as Tucker, while Kerrie Tucker is described as Kerrie.)

1

Tinkering

It just so happened that when the Global Financial Crisis hit, I started analysing a sector of people who were immune to its worst assaults. Over several years I came to know these people as tinkerers. But explaining tinkerers in a methodical way became hopelessly reductionist, and also too expansive – like over-explaining a joke.

I felt like the square-kid trying to codify the cool-kids' fun. And it seemed phony, researching a doctoral thesis at Swinburne's Institute for Social Research, where my colleagues investigated weighty topics about social justice, civil liberties, human rights, digital communities, legal frontiers and urban sustainability. I hadn't yet learned that this ineffable thing called tinkering had its roots and branches in every one of these fields.

This is in part because my own piecemeal 'career' (nanny, waitress, dish-pig, barmaid, graphic designer, outback camp-cook, farm-hand, journalist, editor) hadn't yet equipped me with the tools to analyse tinkering culture. Even so, those of us who've never had a straight career trajectory, but have largely dabbled from home – makeshifting and free-ranging from one vocational direction to the next, acquiring odd skillsets and eking out a freelance living along the way – we recognise some facets of the tinkering mindset. To First World people lucky enough to accrue skills and live opportunistically, secure enough with uncertain outcomes and resistant to labour-market impositions, the tinkering mindset described in this book might seem so obvious as to be laughable.

I'm guessing this is why a few tinkerers poked fun at me when I started asking questions. It wasn't because they themselves lacked a scholarly curiosity. Some tinkerers I met (John D'Alton, Eddie Banks, Gilda Civitico and her partner Andrew Peel) had or were completing their own PhDs in science, engineering and the humanities. Despite what could be described as their manual 'work-choices' (an Orwellian misnomer in Australian political culture), many tinkerers I studied were tertiary-educated, and all were passionate polymaths and autodidacts. Tinkering itself is a form of scholarship.

But some had an attitude that might be interpreted as anti-intellectual. This book's thought-whisperer, the formidably talented South Australian tinkerer Mark Thomson, was especially fond of poking fun at my academic inquiries. (In turn, I've nicked from Thomson many of this book's ideas – the term 'tinkering mindset' is one of many cadged off him.) More than once when I asked about his tinkering practice, he teased me. '*Practice!*' he scoffed. He told me he regarded that word as the language of arts-administrators, a contemptible bullshit-profession. *Practice* wasn't the vernacular, he said, of everyday tinkerers tooling with engines and fixing up cars in their sheds. Those tinkerers don't self-consciously consider their *practice*; they don't engage in self-important careerist posturing with fashionable artists' statements to impress creative-industries bureaucrats and credulous arts-consumers. (Or something like that.)

In many ways, Thomson reminds me of the late art scholar and philosopher Peter Dormer, who complained that contemporary artists are pampered and encouraged to 'seek the apotheosis of uselessness', while skilled everyday makers moved in 'a world of modest ideas with a straightforward vocabulary of familiar and functional forms'. Himself tertiary-trained in fine arts, Thomson told me that some tinkerers he'd met (he's author of *Blokes & Sheds*) even shun the term 'creative' because of its effete, self-serious inferences, but that

Cover detail of the satirical *Wondermark's Tinkerers' Handbook,* by David Malki ! ("I spell my name with an exclamation point"), available at wondermark.com. David Malki ! offers his comics to be used permission-free.

he'd seen enough to understand that putative 'non-creative' manual enterprise (like engineering and smithing) usually involves expansive imaginary, inventive and expressive powers that trump many of those you might find in public galleries.

One of the many insights to be mined from Thomson's perspective is that domestic tinkering in Australia is often experienced as a tacit habit that resists easy explanation or category. To many everyday tinkerers – people who habitually devote their time and passions to adapt, invent, mend, create, modify, repurpose, renovate, improvise and build from their homes – tinkering is unremarkable, ordinary, obvious, common-sense living.

But when you cast a scholar's gaze upon the tinkering life and its products, tinkerers emerge as remarkable, and worthy of deep discussion. The tinkering life isn't for everyone, but committed tinkerers have much to teach all of us about negotiating power and living well, especially amid current conditions of job insecurity, wage disparity, career precarity, globalisation, privatisation, housing unaffordability and deskilling in all sectors – and the disintegrative impacts these can have on our livelihoods, identities and humanity. In other words, especially under the dehumanising effects of late neoliberalism.

Many of us are tinkerers, but some of us are more tinkerly than others. This book is about people with an extraordinary commitment to tinkering: people who habitually prioritise producing, salvaging, repairing, inventing and adapting things above consuming or commercialising things. These people are do-it-yourself (DIY), make-it-yourself (MIY) and do-it-together (DIT) practitioners who experience tinkering not as avocation, but as vocation. In other words, tinkering at home is their life-career – even as it serves as leisure. Though some of these tinkerers have formal work commitments (commissions, part-time jobs), their life-focus isn't on external work, but on informal home production.

Tinkering

For a sector of Australians, tinkering is an impulsive habit of material problem-solving with its own lexicon of values. It's an iteration of freedom, pride, dignity, ethics and artisanal joy – an unfettered way to live according to one's own measures. It can be practical and utilitarian, a form of economic production, but also a form of scholarship, play, adventure, resourcefulness and resilience. It can be a portal to social connection, community, spirituality, sanctuary, thrift, identity and political resistance.

Who knew? How can a commitment to pleasurable life at home be these things? You might choose to make your clothes, fix your washing machine, build a staircase, devise a prank machine, invent your biofuel processor, improvise your recipes, make a spycam, produce a hybrid vegetable and weld your front gate – but how can such material adventures be *methods* to resist state and market impositions? How can such a life be economically viable? How can this life count as leisure and work concurrently? How can it be consumption and production, individualist and communal, conservative and progressive, amateur and professional, physical yet metaphysical? The tinkering described in this book ridicules so many of the taxonomic binaries we take for granted as the natural order of things. Sometimes these binaries entrap us, channelling our lives into repressive corners with little room to move, and by revising them using the tinkerer's mindset, we can reimagine the way we live.

Although domestic tinkering is steeped in ideals of autonomy and self-reliance, its effects run deeper than dilettantish individual practice. The French theorist Henri Lefevbre wrote about the subversive potential of everyday life, contending that certain forms of leisure carry the seeds of revolutionary social change. In tinkerers' lives, the spheres of home and work are no longer cleaved. When leisure at home also happens to be your work, you're engaging in a broad political act. You're not merely 'opting out' or 'downshifting' or even 'working from home'. On the other hand, nor are you being overtly political and effectual in the way movements like Black Lives Matter are. But you're actively rebuffing modernity's deepest social structures and contracts, in large part because your life is no longer split between active work and passive leisure. The work–leisure fragmentation of your selfhood is reintegrated. To the people profiled in this book, leisure isn't a compensatory break from work – instead it's a highly-productive and pleasurable career, and, as such, it's a source of self-worth, security and fulfilment not directed by the market or state. The broad social consequences of this can be unintentional, but they can also be profound.

This book is written at a moment when a populist fury has erupted across developed economies – a force attributed to growing economic inequality and concentration of wealth and power. But this fury is also fuelled by obsolescent ideals of work, home, identity and belonging. Many populations within these economies are failing on almost every metric that psychologists and sociologists use to measure happiness, even as they enjoy their relative wealth and freedoms during peacetime. Against this backdrop, tinkering isn't a counter-cultural movement as such, in part because tinkerers themselves aren't always an organised collective. The people featured in this book are united only by their values and habits. Everyday tinkerers are very often individuals and collectives of people who work at home according to artisanal and amateur values: agency, freedom, risk, experimentation, originality, learning, play and autonomy. Within a culture of precarious social and labour conditions, these people manage to tend to their senses of wholeness.

Tinkerers' stories show how a sense of agency in the world is generated from this holistic base. They show how tinkering's values of production themselves foster production of values. These values circulate through our culture, materially and ideally. By focusing on tinkerers, we can understand many ways that people negotiate modernity's impositions; or the impositions of postmodernity, or late modernity. All these terms are flashpoints for scholarly quarrelling, but 'modernity' serves in this book as a heuristic catchall to grapple with the generalised conditions of contemporary Australian – and globalised – life.

Spanning across disciplinary boundaries, this book is an attempt to go some way towards explaining those cool-kids' fun by using square-kid frameworks invented by some impressive scholarly minds. Sometimes there are mismatches between academic language and lived experience, but I'm hoping this book might correspond with both a scholarly and a more general-purpose readership. I hope that by framing tinkerers' stories using conceptual tools forged by some masterful theorists, it can go a little way towards helping people (especially policy makers, and more especially tinkerers themselves) recognise why ordinary domestic tinkering is more culturally, socially and economically important than the paucity of research on it might suggest.

Stories in this book reveal tinkering as both a cultural resource and a cultural force that tests and negotiates power structures. They also show how tinkered products embody political and cultural values that circulate through informal economies. A list of those products – preserves, spycams, jewellery, soft-toys, biofuel processors, artwork, music-boxes, cars, robotics, bike contraptions, house-builds, moonshine liquor – might seem wildly indiscriminate, but tinkering isn't a discrete field of material production. Instead, tinkering – like craft – is a set of concerns implicated across many genres of production. (I have the craft historian Glenn Adamson to thank for this insight.) So this book's focus isn't solely on the products of tinkering, fascinating though these are. It's on the cultural ways tinkering practice is invented, reinvented and circulated. By focusing attention on tinkering as a locus for discussion, we can gain insights into the profound value of everyday informal cultural production.

Do people really regard themselves as tinkerers? What kind of people do this? And what exactly is tinkering? I've already described it as a 'set of concerns' and a 'locus for discussion', but is it really a thing? In 2009, as the GFC hit, I searched for a non-fiction book on the subject of *tinkering* or *tinkerer* at Amazon.com. This search was motivated by a convergence of ideas and advents. One was my own vague observation that an assortment of commentators – journalists, linguists, cultural studies scholars, lawyers, science sociologists – had been grappling with the power struggles between those of us who use consumer gadgets and appliances, and the powerful corporates who manufacture and regulate them. Many news stories emerged about technology and its discontents: IT geeks conspiring to hack Microsoft's 'unhackable' Xbox console; new cars whose sealed engines resisted home-maintenance; laptops and handsets that rapidly became obsolete; mobile phones changing the ways we behave; drones intruding on neighbourhood privacy; a cancer cluster among ABC Brisbane staff who worked under broadcast towers; another in Melbourne among RMIT staff working beneath a mobile phone mast; a Sydney man who stole a vintage army tank and rammed it into seven phone towers, disabling six. There were complaints about IT help-desks and call-centres being outsourced to distant countries; claims of illness around wind turbines; university lecturers using signal-scramblers to block their students' mobile devices; parents' objections to the rolling out of school wi-fi hubs that had been banned in Europe; agritech multinationals threatening our food sovereignty – the list seemed endless. In newspaper op-eds and radio broadcasts, commentators were discussing the extent to which product design and public policy were being framed in the interests of those of us who use, assemble, or bear the impacts of these technologies.

Usually in these discussions (and in my own thinking), it was simply taken for granted that policy – and by extension, product design – was framed by a distant sector of industry bodies and public-servant brokers variously called 'bureaucracies', 'technocratic elites', or what sociologist Sheldon Krimsky called 'the industry–government complex [that] ignores the voice of its own citizenry'. Many of us who cursed our computers and shared bitter stories about Telstra and Optus felt estranged from the everyday tools that were our primary means of connecting with people. We didn't (and don't) deeply engage with or understand these tools, let alone the industry–government complex that designed, regulated and powered them.

But this wasn't an inevitable condition of modernity. It was a designed one. It didn't happen because the technology was complex, but because of cultural and political intent: regulatory restrictions such as shrink-wrap clauses; licence contracts and warranty waivers that prohibit us from opening our own machines; product design restrictions such as non-replaceable parts; planned obsolescence; distant bureaucracies and self-regulating industries whose public engagement is systemically stonewalling. All these things. It was only a few decades ago that consumer-technologies contained step-by-step instructions for home repair and adaptation. Now, they contain legal threats against these.

We became estranged from our basic kitchen tools, too, as even the most straightforward human task of preparing food was mystified. Marketing their products as convenient, corporate marketers were promoting a culture of learned helplessness. Manufacturers started telling Australians that even the most basic culinary practices – steaming rice, popping popcorn, or brewing coffee – required skills too specialised and troublesome for a layperson to manage. They invented microwave popcorn, 'instant' rice, and standardised coffee pods that can't be composted (there's a growing industry of nylon tea-bags, too).

These industries flogged expensive high-end gadgets like Thermomixers, so we no longer needed to use our muscular or sensory judgment when following recipes. Even fresh fruit and vegetables were being spruiked in ways that presented them as too difficult for the layperson to manage and understand, seeding a market for expensive biotech foods we don't (or shouldn't) need, such as non-browning potatoes or avocados, and fortified cereals. Food sociologists started observing that as ratings for television cooking shows soared, their audiences, inversely, became disengaged from everyday food literacy. Celebrity chefs were mystifying food as they turned cooking into elite competitive sport, while lending their names (or brands) to supermarket convenience food, actively dumbing-down our everyday competence. The social impacts of this – decreased First World food security, poor public health and loss of cultural custom – are profound.

As I started this research, two best-selling books – US philosopher Matthew Crawford's *Shop Class as Soulcraft* (2009) and UK sociologist Richard Sennett's *The Craftsman* (2008) – provoked discussion about the individual and social value of manual competence and material literacy in post-industrial economies. These books mounted a moral argument for our reacquaintance with manual skills and material (technical) knowledge within our deskilled cultures of so-called information economies. Sennett's and Crawford's arguments resonated strongly with other clusters of discussion and discontent, and the growing popularity of slow lifestyles, localism, craft revivals and artisanal production.

The folkways and values of bygone times tend to be discovered anew by every generation, and by 2009 – in the wake of the GFC – books that ushered us backward in time, slower in pace, more human in scale and closer in proximity had amassed into something resembling a movement. In Australia, Patrice Newell's *The Olive Grove* (2000), a downshifting tree-change memoir and political inquiry, was followed by David and Gerda Foster's *A Year of Slow* (2002), a diary of the couple's return to the cycle of planting, eating and living simply with eight children (with recipes thrown in). Clive Hamilton's *Growth Fetish* (2003), *Affluenza* (2005) and *The Freedom Paradox* (2008) were all concerned with the ethics and excesses of modernity and the values of late industrialised capital. Anne Manne's *Motherhood* (2005) suggested a return to traditional family values (not in a reactionary, anti-feminist sense, but in a takes-a-village-to-raise-a-child sense). David Potts's *The Myth of the Great Depression* (2006) identified apparently positive social aspects of the Depression years; Jeff Sparrow's *Communism: A Love Story* (2007) mourned Australians' loss of passion and political idealism; Germaine Greer's *Whitefella Jump Up* (2003) asked Australians to purge ourselves of cultural malaise by returning to the point where we went wrong to acknowledge our shared aboriginality. In the US, titles from *Deep Ecology* (2001) to *Fast Food Nation* (2001) and such books as *Supersize Me* (2004) and *Reset: How This Crisis Can Restore Our Values and Renew America* (2009) warned us to rethink our consumption and return to lower-tech, less industrialised, smaller-scale solutions. Michael Pollan's internationally bestselling *The Omnivore's Dilemma* (2006) and *In Defence of Food* (2008), based on groundwork laid by University of Melbourne food sociologist Gyorgy Scrinis, urged us to return to traditional and local food customs, from paddock to plate.

From Joseph Stiglitz's *Globalization and its Discontents* (2002) to Bill McKibben's celebrated *Deep Economy* (2007), we learned the ways that globalised industrial capital is eroding humanity and the earth's finite natural resources.

Such literature seemed to have peaked, and I wondered if this signified a cyclical neo-Luddite moment, redolent of other such moments in my lifetime – championed by 1970s public intellectuals such as E F Schumacher and Wendell Berry, and later by 1990s figures including Jerry Mander, Langdon Winner and Bill McKibbon. Despite Australians having strong trust in science, and despite us being famously early adopters of new consumer technologies, by 2009 there was mounting evidence that many of us were uneasy about the regulations surrounding these, and suspicious of the benefits touted for them. Studies in Swinburne's *National Technology and Society Monitor* consistently found that most Australians didn't have much trust in government institutions or major companies when it came to information about many technical devices or systems we're exposed to.

Envisaging a research topic concerned with how some people maintained a sense of agency over everyday products, I started interviewing DIY technologists – people who modified, assembled and understood gadgets and processes, or people who rejected technologies they couldn't understand, fix or modify. I started by interviewing people within the then-nascent steampunk movement, a DIY culture I describe in Chapters 6 and 8. I began by observing their practices and asking them about their material behaviours. My interviewees tended to describe their DIY practice as *tinkering*, some even describing themselves as *tinkerers*, and although for years I ignored this word (it seemed cliquish and insignificant), I eventually noticed that media reports and broadcasts were also discussing a vague dilettantish practice they called *tinkering*.

In these media reports, tinkering was characterised as a *method* that was said to be giving people a sense of agency over everyday technologies and systems. It was reportedly something that occurred in suburban backyards, sheds and kitchens, but which had great empowering, democratising and revolutionary potential. On ABC Radio National, futurist Alex Pang spoke of 'tinkered solutions that encourage people to do things that governments have had a really tough time doing … through either regulations or taxes … probably we're going to end up with a combination of a billion little tinkered approaches to things like energy savings or water savings'. Pang told listeners tinkering was a 'powerful' means of individual freedom, encouraging:

> a kind of spirit of open-ended inquiry about the material world, and it also encourages a kind of empowerment, a kind of sense that you don't just have to take the world or take things as they come, but rather that even in today's apparently high-tech, slick, very pre-processed, very produced world, it's possible still to break open the covers, to get into the silicon, to get into the gears and to improve them, and to make them your own …

Also on ABC Radio National, Program Director of Industrial Design at the University of New South Wales, Miles Park, described this tinkering as 'a decentralised activity' that was 'all about appropriating things that are to hand, it's resourceful, it's economical, it's opportunistic … it's very much about the top-down meeting bottom-up'. Park characterised tinkering as a way to resolve environmental problems in an Australian culture (or policy climate) that doesn't require product stewardship. In other words, there is little policy governing the responsible life-cycle of products:

> So an awful lot of this stuff is going to landfill, huge waste of resources, huge amount of embodied energy in these products

> that's being lost. So … there is a definite connection between the life-span of products and tinkering.

In online sites such as TED and YouTube, others were making similar claims. Deloitte scientist John Seely-Brown described tinkering as a way to 'play with knowledge, to play with creating knowledge … to create knowledge on the fly by experimenting with things'. And, similarly, Pang said that tinkering was:

> [not] so much a specific set of technical skills: there tends to be a pretty instrumental view of knowledge. You pick up just enough knowledge about electronics, textiles, metals, programming, or paper-folding to figure out how to do what you want. It certainly respects skill, but skills are a means, not an end: mastery isn't the point, as it is for professionals.

Also at this time, Mark Thomson from Adelaide's Institute of Backyard Studies (IBYS) told ABC Radio National listeners that tinkering was primarily about self-reliance, but it had the potential to address such colossal political problems as climate change:

> Tinkering has a strong tradition in Australia as a result of particular conditions here. I mean, primarily it was isolation, you know, that when white people first came out from Europe, there were not enough resources to be able to make things. If the plough broke, you couldn't wait two and a half years for a new one to come from Birmingham. So you tinkered around, you adapted, you found all sorts of resourceful ways of fixing things up. And there were things like the Depression, wars, a lot of human causes, but primarily the causes for that resourceful tinkering were environmental, they were fire, floods, drought, and they have driven the sense of tinkering very powerfully. And I actually think that climate change will be the greatest challenge for the tinkerer. And I actually think that the small end of town will respond better than the bigger end of town.

And beyond Australia, popular magazines such as *Slate* and *Wired* were claiming tinkerers' unorthodox engagements with consumer technologies was inspiring the decline of planned obsolescence and the extension of product life-cycles.

Yet for all these claims, there seemed no deep analysis of how tinkering performed its (ostensible) revolutionary feats. This is why, having exhausted scholarly searches, in 2009 I entered 'tinkerer' and then 'tinkering' into Amazon's search engine. By this time, Microsoft had already launched *Tinker*, its robot puzzle software, which suggested that the idea already had strong cultural currency. This wasn't reflected in my Amazon search. It threw up a single non-fiction title, historian Kathleen Franz's *Tinkering: Consumers Reinvent the Early Automobile* (2005), plus a handful of fiction titles from various eras. Now, as I write this, this single figure has ballooned to a whopping 480 results for the same search. While this may be partly explained by updated search-engine algorithms, there's no question that a formidable boom in DIY literature is upon us. Fiction and non-fiction alike, this literature champions the advent of the *maker movement*, a widely-reported phenomenon originating from North Hollywood's hugely successful *Make* magazine and its worldwide Maker Faires. To American commentators at least, tinkering is a defining feature of the maker movement (discussed in Chapters 2 and 8).

Did this cultural conversation have anything to do with what's happening in Australia's kitchens, paddocks, sheds, studios, workrooms and backyards? There was no scholarly study available that answered this question. In his 2008 book *Serious Leisure*, sociologist Robert A Stebbins observed that 'No systematic research yet exists' of tinkering; and that 'tinkering pastimes remain understudied'.

But it wasn't the cultural or scholarly conversation that gave me this topic. Instead, while I was looking the other way, my research participants colluded within the pages of my notebooks, building and modifying the topic even as I was theorising their stories in other directions. Failing to attract my attention, they eventually bludgeoned me over the head with it. 'One never knows what a book is about until it's too late,' wrote William Mitchell, after his 1994 book *Picture Theory*, once released into the world, declared itself to be something different to what he supposed. Mitchell first realised this when a reviewer proclaimed that *Picture Theory* should instead be *What Do Pictures Want?* The current edition is revised, and carries this title.

A similar thing happened here. This project was once called *The New Luddites*, among other ham-fisted titles. (I'm indebted to Ramon Lobato for disabusing me of these.) Initially, 32 people were interviewed for my study of DIY and DIT practitioners. Although these 32 people have a presence in the stories that follow, only seven ended up as this book's protagonists. I didn't know these seven were 'tinkerers'. I chose them simply because they were able to articulate their DIY stories well, and because I liked and admired them.

Liking and admiring seven people might seem an absurdly arbitrary and sloppy method to generate rigorous 'data' (a word ludicrously inadequate for tinkerers' stories). The sample seems too small to be meaningful, and rapport as a method of data collection would certainly have built-in blinkers. Wouldn't *liking people* make critical distance – one benchmark of scholarly rigour – trickier? How can you possibly produce detached analysis when you so strongly identify with your research subjects? (*Identify* is too dispassionate a word: I'm enchanted and smitten with each tinkerer-protagonist, and most have become firm friends.) Isn't there a certain danger of producing hagiography instead of analysis? Surely you're just cruising in your comfort zone, affirming your beliefs, and avoiding those discomforting tensions between the interloper-researcher and the repugnant-other that can make for the edgiest findings?

These are questions that haunted my research, but in fact, rapport as a qualitative research method has a strong scholarly lineage – albeit one that's suffered rancorous argument. Within anthropology traditions, rapport with participants is regarded in wildly contradictory ways: from gold-standard research to data-by-stealth. If you try to follow this strange debate over time, it amounts to something vaguely resembling good-cop-versus-bad-cop research. On the one hand, rapport is considered a central fixture in ethnographic research – a legitimate strategy to generate unguarded, relaxed responses (good cop). On the other, it's considered a misguided catch-all attempt to clean up the mess of subjective fieldwork – an uncritical way to rationalise the irrational (bad cop). Either way, your goal is to extract insider-confessions that will resolve your investigation.

The stories in this book aren't strictly ethnographies, but participant-observation is an ethnographic tool I've used. It essentially involves conversing and participating in tinkerers' projects. (Media academics like to call this 'immersion journalism'.) Early in the piece I wrestled with what has become a clichéd insider–outsider tension in ethnographic research: becoming anxious about maintaining 'professional' or critical distance, but not wishing to impose formalities that upset rapport. Scholars who work outside the humanities sometimes think the 'observer effects' of this type of research can contaminate your data (if data is ever pure), but there are many lovely stories that refute this idea. Some come from sociologists Torin Monahan and Jill Fisher. The best insights, they write, can come from 'cultivating close ties with others and dispelling the illusion that robust data is best achieved through distance'. They describe an instance when they were studying how informed consent worked in the doctor–patient relationship. A doctor with whom they'd established rapport was entirely comfortable with the researchers (Monahan and Fisher) closely observing him during consultations. During one consultation, even though the doctor explained to a patient that a trial-treatment would not improve her condition, he prescribed it to her anyway, because this placated her anxious son (who was there asking questions). The researchers interpreted this as a situation that had become so normalised in medical care that any 'performance' for them wasn't even considered by the doctor. Their findings were: 'The physician being on his best behaviour could not mitigate the risk of potential subjects' misunderstanding of clinical trials.' So observer effects were integrated into their findings.

When I showed up at tinkerers' homes wearing stained overalls and a tool-belt, or in a kitchen apron and bare-feet during a heat-wave, I don't imagine they 'performed' for me or were on their 'best behaviour' (they were too busy nail-gunning the floorboards or explaining how a slider-crank works). Not any more than you'd generally behave and perform for a social visitor. But there are subtle observer effects – theirs and mine – made explicit in the stories that follow. These stories lay bare their own construction. In *Fieldwork* (2015), ethnographer Lisa Breglia mounts a strong moral argument against formalities in research and writing. She cites a mildly condescending passage from an influential 1972 book, Morris Freilich's *Toward a Formalisation of Fieldwork*. It seems to support an idea that researchers conspire with their participants to pretend there are such things as smokos during the work of gathering data:

> At certain times, and in certain types of situations, the anthropologist and some natives find it mutually advantageous to make believe that the anthropologist is not working, that he is indeed taking a break … The anthropologist often finds himself in a dilemma: to achieve his research goals he constantly must be collecting data. However, in many situations it is extremely embarrassing to appear to be working. The anthropologist,

> like the natives, is ready to grasp opportunities to make such inherently embarrassing situations more pleasant.

Oh, the unpleasantness of research into human-folk! Research for this book was fun, and it was informal, friendly and invigorating (like tinkering, it served as pleasure as well as work). But as much as I'd like to poke fun at Freilich's dated and sexist account, any qualitative researcher would recognise the awkward moments he describes. Asking people to open their homes, answer intrusive questions, work physically with you, tolerate your challenges to their views, teach you skills and involve you in their home projects and routines – it's an extrovert's calling. It's not for someone who's by nature semi-reclusive and more comfortable mining secondary sources than meeting strangers.

Even worse is the conceit of assigning yourself the privilege of writing other people's stories. There's a vast library of scholarly hand-wringing about the ethics of representing research participants. Within it, a stand-out essay nails some concerns I experienced with tinkerers. In 'Writing Shame', Sydney University cultural studies professor Elspeth Probyn limns the sense of being an academic imposter, especially when you invent a research topic. '*Imposture*,' writes Probyn, 'implies making it up, hiding behind a wall of competence'. Imposture is also close to *sham*, a cousin of *shame.* Sham and imposture, as well as forgery and fakery, have been central concerns of tinkering practice throughout history (discussed in chapters that follow).

And the fake-it-'til-you-make-it ethos of writing is also indistinguishable from that of material tinkering. Shame moderates the sham in the craft of writing and tinkering alike – the contingent trade-offs in its construction, and the tension between the tricks it pulls and their practical implications. These tensions are heightened by what Probyn describes as a 'shame in being highly interested in something and unable to convey it to others, to evoke the same degree of interest'. This is the type of shame that drives researchers to craft better, more engaging writing, so they might seduce indifferent readers. But the interests of crafting cohesive stories can also compromise your material. (I wince to think of this book's protagonists reading about themselves under the chapter themes and tropes I've assigned to their stories in the interests of a workable narrative.) Like a long lineage of thinkers before her, Probyn sees shame as integral to ethical craft. A private shame can keep you honest. It's a driving force behind good craftsmanship, because shame involves 'the terror of being equal to the interest' of your material; and shame 'forces us to reflect continually on the implications of our writing'.

The inhibitive shame of academic imposture can also distract you from what's going on in the field. When I fell in awe with this book's protagonists, I could only hope their stories amounted to a research topic. I didn't know these seven people as tinkerers. I knew them as DIY practitioners with whom I'd established rapport. They had values in common, and when I told them stories about each other they seemed to concur. There was a certain resonance I was picking up. But they weren't a representative sample of any particular culture through which I could make generalised findings. There's no union of tinkerers (save for the semi-serious Institute for Backyard Studies and odd Hackerspaces in urban settings); they didn't come from any collective, and I had no word to name them, no neat theory in which to frame them. I'd sourced and recruited them with snowball recruitment, a routine tactic in journalism, sociology, marketing and ethnography. This recruitment relies on a chain of contact through informal and formal networks. My provisional research focus (DIY at home; neo-Luddite values) was explained to each of my original sample of 32 people, and the hope was that, as my research focus was refined, the ongoing participants would provoke deeper conversations and insights.

What was I going to do with all these people and their stories? Like 'rapport', 'hope' might seem an absurdly undisciplined research method. But it's the opposite: it requires sustained fortitude and skill. And again, this method has an impressive canon of scholarly precedent. Building on a particular lineage of philosophers (including Immanuel Kant, Ernst Bloch, Walter Benjamin and Richard Rorty), anthropologist Hirokazu Miyazaki defined a 'method of hope' as a responsive rather than merely interpretive method. It's a method whose ethics reorient the purpose of knowledge. Using this method, the researcher *replicates* her participants' hopes (Miyazaki describes this replication as 'a performative inheritance of hope'). This replication works as both as research strategy and as purposive form of knowledge-production. It's almost a missionary approach. Anthropologist Tim Ingold describes it as a means by which the researcher becomes less a mere gatherer and transmitter of stories, and more a *correspondent*, generating stories 'not in order to accumulate more and more information about the world, but to better correspond with it'. To Miyazaki, a method of hope produces hopeful reflections of knowledge, rather than mere interpretations of knowledge. So if hope is a moral obligation (as playwright Tony Kushner has said), method is also 'about the kinds of people we want to be … and about how we should live … Method goes with work, and ways of working, and ways of being,' according to sociologist John Law. And stories in this book reveal how tinkering is itself a method of hope.

Like mindful meditation, or a commitment to a slow life, this method is simple but not easy. At first, I failed it. Three part-time years into this research, the pressure to hone my topic had pushed it in the wrong direction. Just as Mitchell's *What Do Pictures Want?* mistakenly entered life as *Picture Theory*, *The New Luddites* was destined to be stillborn. My flawed hypothesis was that my research participants were consciously practising a form of neo-Luddism – an intentional, political attempt to regain agency over industrialised technologies through DIY practices. But eventually it became clear that this wasn't something they were telling me (or at least, not all of them, or not always). Instead, they were like the Amish of Iowa and Ohio, who are 'ingenious hackers and tinkerers and do-it-yourselfers', according to technology historian Kevin Kelly. These people routinely reject some new technologies but are early adopters of others. (Several psychology journals report that the Amish experience dramatically lower rates of unipolar depression compared with the general US population – in part, it's theorised, because they've screened out the more alienating aspects of modernity.) Likewise, the 32 people I initially interviewed had an opportunistic and selective uptake of various technologies according to their own terms and values, born of their own understandings of materiality, wellbeing, utility, convenience, ethics and fun. So not always from broader political intent; sometimes their motivations were simple pleasure and pride.

So my whole premise was bollocks. I was Monahan and Fisher's doctor, attempting to perform for my academic peers while prescribing misfit treatments toward my subject. My then-supervisors, Julian Thomas and Klaus Neumann, offered sage advice: 'Trust your material.' This is a method of hope, and as it turned out, the tinkerers in my notebooks and audio recordings had shaped the narrative all along. When I paid attention, the topic emerged into view like a polaroid snap, and it started hollering and pummeling me from all directions – really, a cinematic climax that filled my senses with tinkering. Listening to recordings, reading for the umpteenth time over my interview notes, continuing my visits and conversations, it became blindingly obvious that most of my participants described 'tinkering' or identified as 'tinkerers': something I'd repeatedly noted but not noticed. This realisation happened to coincide with tinkering becoming a fashionable topic on radio and an ongoing meme in online media. So tinkering was a thing! Eureka.

On a hot day in 2013, an unexpected foil entered the story, and, like a good Luddite, he threw a spanner in the cogs. Albrun Lobb was a radio gadgeteer from rural Victoria whom I first interviewed long after establishing relationships with other research participants. In the interests of de-identification, there are a couple of details I won't reveal about Lobb, starting with his real name (though he signed a consent form using it). I met him by chance some years into research, in his freelance capacity as a pest exterminator. Large swarms of European wasps had set up shop in my roof cavity. I don't mind living with wasps; nor do I mind exterminating them – but I was persuaded by an anxious houseguest to call Lobb, who was listed in our rural services directory. When he arrived, I offered him a cup of coffee, which he declined until after he finished fuming the wasps, whereupon he spoke abundantly on topics ranging from the insecticide he favoured (synthetic pyrethrum), his science knowledge (self-taught but extensive), his views on the Gillard government's carbon tax (bad, really bad!), people's views towards radiation hazards (ignorant!), and his passion for restoring, repairing and inventing radio gadgets (unbounded). It was here that my interest leapt beyond polite tolerance – I hadn't managed to recruit a radio tinkerer let alone a climate skeptic tinkerer – and I told him about the focus of my PhD. He agreed to an interview.

Lobb has been a most valuable and fascinating informant, and his character plays a minor role in a couple of this book's chapters, but I didn't establish rapport with him. Nor did I establish rapport with his projects; nor his home. Nor did he use or accept the tinkering vernacular (though he signed a consent form which used it). When discussing his gadgeteering I referred to 'tinkering' and he bristled, telling me the term bothered him. Tinkering, he said, denotes useless dabbling; not the skilled activity he clearly engaged in. His material literacy, his DIY mindset and his relationship with his home seemed to align with those of other participants, and so I explained to him contemporary ideas of tinkering. He replied dismissively that he uses the *Oxford* dictionary as his source of authority, and he would look it up now if his edition hadn't been destroyed in the 2009 bushfires. 'Google it,' I suggested.

Leaving his home, I was hit with anxieties about what made my seven protagonists 'tinkerers' while Lobb, who reported almost identical experiences, rejected the term (or idea, or paradigm, or phenomenon). Was tinkering *really* a thing? Or had I just colluded with marketers and hipsters (if such a tribe really exists) to make it up? (As I write, a café called Tinker has opened in High Street, Northcote, Melbourne's fauxhemian centre.) Did an apparent resurgence of tinkering signify anything other than a cliquish or modish meme of limited value? Or, on the other hand, was Lobb a feral exception to a significant movement?

Weeks later, Lobb and I crossed paths at a small-town bakery, and he immediately told me he'd looked up a current edition of the *Oxford* at the library. 'I was right,' he said. Tinkering, he told me, was defined precisely as he understood it. I replied that my research was concerned more with *lingua franca* usage and that perhaps dictionaries will soon catch up; he dismissed this with: 'I get my information from the authoritative source. You can't go higher than the *Oxford*.'

Although there's a postscript to this story that involves Lobb phoning me and embracing the term with fervour (he'd been acculturated), this kind of knowledge-claim, it seemed at the time, was antithetical to the tinkering mindset of play, reinvention and anti-authoritarianism. And my encounters with Lobb also raised questions about the shared understandings inherent in my methods. Tinkerers are people like me, and people I hope to be. They're my tribe. In oddly officious language, the renowned anthropologist George Marcus described this kind of rapport with research participants as:

> primarily a technique among techniques, a means to ends that are completely within the investigator's (and his or her discipline's) realm of definition, purpose, and authority. The attainment of a 'working' level of trust, sufficient to open access to sources of information and data that serve the inquiry, is the goal of rapport.

But my rapport wasn't a *goal* – it wasn't something I *calculated*. It was something I *tended* when it just so happened. I think it was a mutually-felt, reciprocal relationship of resonance. Consensual, but not voluntary. As I've said, I simply fell in awe – love, even – with my protagonists, their projects and their homes. Peculiar as this may seem, this love involved their material sensibilities and the objects with which they surrounded themselves. And so the stories that follow explicitly acknowledge the agency of objects as well as people, a premise famously articulated in actor–network theory (ANT), now generally taken for granted in cultural studies. To political theorist Jane Bennett, material objects have 'productive power of their own'. They hold 'the ability to make things happen, to produce effects' within their 'agency of assemblages'. Objects, she explains, are players in a vast distribution of agency that spans across human and non-human forces. (If this sounds like crackpot malarkey, try bolting through a revolving door to chase a tennis-ball.) Protagonists' homes and projects seduced me with their atmospheric charge. In cultural studies language, they had 'deep affective agency'; which is to say, these things hit me in the heart.

I tinker, therefore I am. Rapport, to repeat Law on method, is mostly about 'a way of being' and 'the kinds of people that we want to be'. (As it turns out, so is tinkering.) For reasons instrumental and personal, this irreducible criterion was used to invite seven tinkerers to be this book's protagonists. All turned out to share similar political outlooks and common understandings of tinkering. Albrun Lobb's resistance to these understandings shows how my method screened out other mindsets, and in doing so *defined* my topic. Methods do create realities, as well as extracting them. My rapport-as-method was subjective, irrational, emotional, impulsive and biased, and as such it unveiled a certain empirical truth – that there is indeed a sector of people, with important things to teach us, who identify as tinkerers.

2

Hindsight

In Christian folklore, *tinkers* 'drove the nails that pierced Jesus' and were cursed 'like the Jews, to be wandering always', according to the American historian Mary Burke. The tinker figure emerging from centuries of European writings about vagabonds was an itinerant worker of Slovak origin who repaired kettles and pots, and also made wire sieves and baskets. The word *tinker* is variously thought to originate in Irish, Gaelic, English or Romani – to *tink* in Romani is to *rivet*; in English *tinker* was a 'worker in tin' in the 13th century. But the word may have an early Germanic pedigree – the German word for tin being *zinn*, the Danish and Dutch being *tin*, and the Swedish being *tenn*. *Tinker* had artisanal associations in Scotland, and diasporic associations in 12th-century Ireland, where *Tynklers* and *Tynkers*, with their own dialect, are thought to precede the Irish ethnic minority now known as Travellers or *an Lucht Siúil*, 'the walking people' who performed underclass tasks such as knackering and pot-repair.

To understand why tinkering's cultural moment is now, it's instructive to tease out why tinkerers have always been changelings. At some cultural moments they are our saviours and idols; at others they are depraved, deviant, treasonous and even murderous. Their names, too, are shape-shifting. Irish travellers who didn't want to be stigmatised with the *Tinker* label recast themselves as *Gypsies*. But then *Gypsy*, too, couldn't avoid stigma. Unlike the 'honest' trades, tinkers were thought to have underhand means of procuring work – and the associated *Gypsy* became yoked to *being gypped*, or swindled or defrauded. In the Luther edition of *Liber Vagatorum* (1509), the earliest known writing on vagrancy, tinkers 'go about full of mischief, and if though givest them nothing, then one of them may hap will break a hole in thy kettle with a stick or a knife to give work to a multitude of others'. In short, they were seen as an organised ring of scammers.

Even so, this 'underworld of organized, cant-speaking, deviant' people had an important role in pre-consumerist Europe, where sumptuary laws were designed to prevent social and geographic mobility. These laws forbade people to spend or consume beyond their station in life, making practices of thrift and repair highly valued. These laws also made yeomanry consumption beyond basic needs 'a form of sin, rebellion and insubordination against the proper order of the world,' according to sociologist Don Slater. But with industrialisation, the project of mass-consumerism demanded freedom of social mobility, and so 'improvement' overtook 'thrift' as a measure of a person's good character. And so the economic role of the tinker was displaced as cheaper materials – enamel, stainless steel, plastic – became the substances for mass-production.

By the 1800s, the tinker as a symbol of the miserable, opportunistic underclass was still evident. Burke documents a tourist to Ireland who observed 'a miserable scene of a tinker mending the utensils of an equally poor woman'. In the 1850s, *tinkering* still had manipulative, cunning, devious and almost supernatural associations, much in the ways *craft* had in previous centuries (as in 'witchcraft' and 'crafty fox'). In his 1677 technical instruction guide, craftsman Joseph Moxton explained that he named the guide *Mechanik Exercises* rather than *Doctrine of Handy-Crafts*, 'because *Hand-Craft* signified *Cunning*, or *Sleight* … which cannot be taught by Words'. Associations between craft, cunning, tinkering and hacking persisted: in Charles Dickens's 1859 *A Tale of Two Cities* the French Revolutionary Madame Doufage secretly encodes within her knitting the names

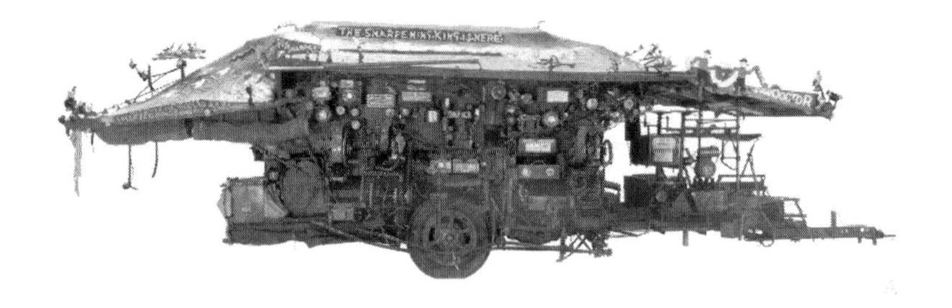

Harold Wright's 9.2 metre Tinker Wagon or Road Urchin (1935).
Photograph: National Museum of Australia.

of people killed, and current discourses surrounding craft-hacking and craftivism also liken digital source-code to knitting technique. A link between encryption, hacking and knitting was still evident during the Second World War, when the British Office of Censorship banned the posting of knitting patterns abroad, fearing they carried coded messages (escape routes were known to be mapped out in quilts). The Belgian resistance also reportedly recruited women to encode activities in their knitting.

This cluster of craft-cunning associations persists: 'spin' remains a byword for devious representation or weasel words in English (as in 'spin a yarn' or 'PR spin'); 'fabricate' denotes both faking and making; and 'patch' still denotes makeshift security protection of vulnerable (or hackable) software. An etymology of 'hack' and 'hacker' denotes on the one hand everyday, mundane utilitarian craft and repair, and, on the other hand, clandestine activities outside the norm. 'Hack' continues to mean to cut roughly, to chop, or even to mutilate; it was for centuries used to describe a worn-out workhorse, and often these days describes someone employed to undertake routine work, especially in journalism. Hence 'hackneyed' came to denote something overused, worn out and unoriginal (as in a 'hackneyed phrase'). But according to MIT historian T F Peterson, 'hacker' originated in the 1950s and 1960s to describe technology pranksters who engaged in clandestine practical jokes to test and demonstrate their level of skill, ingenuity and imagination in diverting technologies from their everyday use. Post-1960, 'hacking' migrated into common usage, describing disruptive and deviant computer hacking. Hackers, according to cultural studies scholar McKenzie Wark, became 'a new kind of folk devil'.

Consider the shifting meanings surrounding *hacking* – tinkering's habitué. As tinkering's meanings have been revised and revived in the past decade, the meaning of hacking has been repurposed in parallel ways. No longer characterised as an act of sabotage or deep subversion, it's now a cool route to entrepreneurial success. In Australia, hacking now has commercial support from the very technocratic complex that criminalised the practice in the first place. Community-run Hackerspaces, for example, are sponsored by CAD and CAM tech companies, while *GovHack* (an 'open data hackathon') was launched in 2009 by the Australian Government, with the sponsorship support of Telstra, HP, Microsoft, and Linus Australia, among many other tech giants and government agencies (including IP Australia, the ATO, the ABS and CSIRO). *GovHack* encourages 'our civic hacker community' to work with 'open source government data' to build 'a better democracy through innovation, participation, and the development of a strong community of civic hackers'.

In Britain and America, too, hacking is shedding its countercultural, activist and outlaw implications and has become thoroughly popularised and inverted. Historically seen as a seditious or treasonous threat, now hacking *upholds* law, order and national security, and is celebrated for the way it 'makes good citizens'. Indeed, Silicon Valley venture capitalist Dan Abelon equated hackers with entrepreneurs when he stated in 2012: 'I am advocating for governments in every country to recognize that the health of their economies will be increasingly dependent on whether there is a decently paved path for hackers to start companies … Nations should consider hackers to be a precious resource.' In the *New Yorker*, technology critic Evgeny Morozov summarises:

> the White House endorsed the first National Day of Civic Hacking. In Britain, the Metropolitan Police … helped organize 'Hack the Police!' – a so-called 'hackathon,' where software developers and designers were encouraged to bring their 'unique talents to the fight against crime.' … 'I'd like to see the spirit of hackerdom improve peace in the Middle East,' the influential technology publisher and investor Tim O'Reilly [publisher of *Make*] proclaimed a couple of years ago.

There's plenty to suggest these developments have less to do with civic engagement than with anxieties around the sabotaging of globalised capital and global security. They're a global extension of the more nationalistic approaches to hacking throughout history.

Consider the Bramah lock controversy from 1851 to 1853, in which American locksmith Alfred C Hobbs fended off accusations of 'tinkering' from the English when he managed to hack into the secrets of the hitherto unhackable English Bramah lock. As with current-day corporate and WikiLeak-style hacking, the Bramah lock hacking issue was framed as treasonous; a threat to national security (an idea that has always been conflated with the interests of private capital).

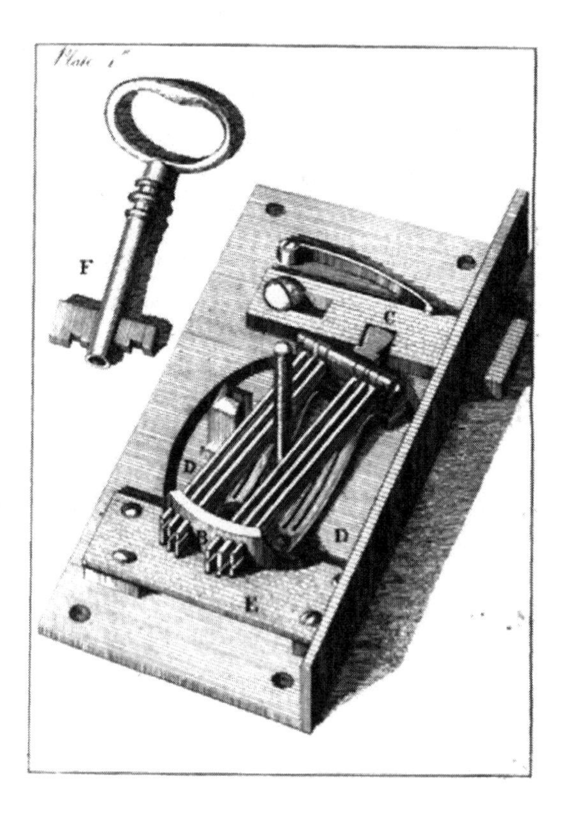

The 1850 'unhackable' English Bramah Lock, successfully picked by American locksmith-hacker Alfred C Hobbs.

The lock was patented, and according to *Cabinet* editor Jeffrey Kastner, 'personal safekeeping was a microcosm of larger political concerns about security … Britain that had used its superior ingenuity to acquire vast wealth also had to be able to effectively protect it.' In an 1853 letter published in *The Times*, Hobbs riposted his English critics, objecting that they 'will hardly catch me "tinkering" – as it is elegantly expressed – "with my basket of instruments," either upon their demand or that of any other man'.

An association between outlaw encryption and tinkering persisted in other ways. According to Kellow Chesney's *The Victorian Underworld* (1970), tinkers were regarded as outcasts whose deceitful and dangerous intent was encoded in a 'cryptic' cant called *Shelta*, the tinkers' language putatively based on Irish Gaelic:

> Almost the only thing that is quite unambiguous about the tinkers is that they are to be counted among the very lowest and roughest of the wanderers. [...] There is a revealing little anecdote of a scholarly observer earnestly taking down notes of Shelta in a low pub when, suddenly, he catches the actual gist of the speakers' intentions and, ramming the table against their legs, bolts for his life ...

The tinkers' *Shelta* was full of the 'inversions and cryptic devices typical of the jargons of the outcast'. Although the valued 'travelling tinkers' were also recorded until as late as the 1960s, the use of *tinker* in subsequent decades was largely confined to literary tropes of magic, trickery, manipulation and hoaxing. (To this day, as the stories in following chapters will show, tinkerers embrace the fake-it-'til-you-make-it ethos.) The pre-Disney character Tinker Bell, originating as a tempestuous, sexual, prankster fairy who mended pots and kettles and 'spoke' a ringing-bell dialect that only other fairies understood, featured in J M Barrie's 1904 play and 1911 novel *Peter and Wendy* (later, *Peter Pan*).

Tinkering is at once estranged from and at the core of dominant Australian culture – and at once a requisite and nuisance for modernity. John D'Alton, a tinkerer profiled in Chapter 8, described tinkering as a 'defining characteristic' of postcolonial Australian senses of identity. 'Particularly,' he said, 'as Australia has only been settled for a couple of hundred years'. By way of example, he referred me to Russel Ward's classic and influential *The Australian Legend* (1958). To this day mined for its populist tropes, *The Australian Legend* depicts the Australian 'character' as a makeshift survivor:

> the 'typical Australian' is a practical man … He is a great improviser, ever willing 'to have a go' at anything … If rough and ready improvisation were convict traits they were also, in the outback, often necessary conditions of survival. Where population was so scattered and specialist service of all kinds practically non-existent, a man had to be a jack-of-all-trades who knew how to make do with what ever scanty materials were at hand.

But *The Australian Legend*'s characterisations have been thoroughly berated. Historian John Hirst, for example, wrote that 'the success of the book became the best support of its claims'; and historian Miriam Dixson argued that Ward's characterisations ignored 'human beings who happened to be female' and by doing so reinforced a dominant patriarchal view that 'threw up a peculiarly limited style of masculinity that owes a lot to that strand of national identity sketched by Ward'. Both historians believed that Ward shaped myths as much as he documented them. But despite Ward's characterisation of the 'typical' Australian as a man, mythologies surrounding 'the bush' continue to exert powerful influence over how Australians are imagined, represented and understood. The heroic 'bush carpenter', an 'unorthodox artisan indeed', was mythologised in Henry Lawson's 1902 story, 'The Darling River', and continues to enjoy strong cultural currency today.

In his epic monograph *The Bush*, Don Watson describes 'resilient settlers' – a frontier Australia that fostered 'the durable belief in improvisation, or "she'll do"'. Cultures that produce their material things under frugal conditions tend to have such words and phrases firmly embedded in their language; parallels of *she'll do* or *she'll be right* are iterated in *chabuduo*, denoting 'close enough' craftsmanship in parts of China; the Hindi word *jugaad* is used in India to denote frugal innovation, as does the Brazilian word *gambiarra*; and in some parts of Vietnam *mày mò* can denote quick fixes and making-do improvisation, as well as more contemplative tinkering.

And Australian tinkering continues to be interpreted by historians within consistent narratives of hardship, displacement and resilience. In *Beyond the Seas*, a 2009 research project examining geographic relocation and its influence on material production, craft historian Stephen Dixon asserts that a tradition of 'uniquely Australian artefacts' with a culture of 'makeshift flair, practicality and eccentricity' developed among settler Australians and their descendants, as well as their displaced Indigenous contemporaries. Convict love-tokens were made of defaced and reinscribed George III cartwheel pennies; the Wagga or bush-rug of Depression years was crudely sewn by men and women from patches of flour sacking; utensils and furniture were crafted from kerosene cans and oil drums; and bush furniture was made from salvaged 1920s oil barrels. Bones from road-kill continue to be incorporated into ceramic artefacts. These are all:

> produced as a direct result of their makers' experiences of cultural and geographical dislocation. These objects are characteristic of an evolving 'making-do' vernacular tradition, influenced by economic hardship, a harsh and unforgiving environment, and a scarcity of resources and familiar materials.

Post-contact, Indigenous Australians incorporated discarded European materials into their making-practices. Broken bottles and porcelain plates were upcycled into spearheads; military gorgets were repurposed as decorative status objects, and organic materials gave way to recycled tin and fencing. More contentiously, settler-Australians continue to appropriate Indigenous artefacts and body parts into a number of ornamental, souveniring and fetishistic contexts.

Beyond the Depression years, the commonplace practice of re-purposing kerosene tins, packing cases, and cotton-reels into home furnishings was promoted in nationalist campaigns as practical and patriotic. Many Australian artefacts made between settlement and Depression years are thought by museum curators to have been crafted not within a vernacular tradition, but purely for utilitarian purposes. Unlike America's entrepreneurial colonial and pilgrim models, Australia's penal colony had a more functional and less sanctified approach to folk-products. Mark Thomson, involved in Maker Faires in Australia and the US, believes American tinkering tends to be more entrepreneurial; whereas Australian tends to be more adaptive and resourceful, although we share the American nostalgia for frontier self-sufficiency.

After industrialisation and with the rise of the middle classes, Australians and Americans shared a deep belief in the moral value of DIY practice. In his 1841 lecture 'Man the Reformer', Ralph Waldo Emerson told the Boston Mechanics Apprentices' Library Association that even among the well-off, 'every man ought to stand in primary relations with the work of the world, ought to do it himself, and not to suffer the accident of having a purse in his pocket'. DIY work, he said, 'is God's education', but the individual who practises it 'by real cunning extorts from Nature its scepter'.

In Australia, tinkered and makeshift objects have historically served as powerful symbols of subversion and folk revolution against displacement and injustice. The Eureka Flag, a national treasure crafted out of repurposed shirts, tea-towels and aprons, remains a radical symbol of rebellion and resilience across the bandwidth of Australian political ideologies. The iconic armour worn by Irish-born bushranger Ned Kelly, crudely hammered out from farming plough mould-boards, is also regarded as a national treasure and seen, according to Dixon, as 'a metaphor of the poor settler society struggling against the casual brutality of the mother country, and the inherent racism and political corruption of the colonial authorities'. And tinkering continues to be idealised as a radical and transformative act against powerful forces, as disclosed in the chapters that follow.

In Edwardian times, itinerant and underclass-labour associations of *tinker* shifted as the term increasingly become a verb associated with leisure activities largely in or from the home, the amateur sphere. Amateur activities, freed from their aristocratic origins since industrialisation, became common pursuits and were accordingly devalued: seen as leisure activities and regarded as by-products of surplus economies. Leisure among the masses, as uncoerced, unregulated and free-time activity, was also seen as a danger in the sense that 'workers will chose more time rather than more goods as the reward for industrial progress', according to sociologist Don Slater. But with the recognition of leisure's potential for nationalistic and commercial projects, tinkering at home became heavily promoted. In *The Design of Everyday Life*, Shove et al. note that the term 'DIY' became evident in US advertising as early as 1912, becoming widespread in the 1950s (as I write, in Australia the domestic hardware and houseware market is worth around $2.1 billion).

Throughout the Depression era, travelling tinkers roamed rural Australia in live-in wagons trying to secure work. In one such 1935 wagon:

> A variety of tools for sharpening and repairing domestic utensils and saws were fitted to the side of the wagon, with a living area at the back. For the next 34 years Harold travelled the length and breadth of eastern Australia earning a living as a travelling tinker and 'saw doctor'.

A *Herald* newspaper portrait of undercover police detective John Christie, taken in Depression-era Victoria, shows him 'disguised as travelling tinker, sitting on box mending furniture' – suggesting that the tinkers of this era were regarded either as an underworld or else a culture through which to gather criminal intelligence.

In C J Dennis's 1917 satirical verse-novella, *The Glugs of Gosh*, the protagonist Sym is a misunderstood tinker who outwits the king and shuns a prestigious life. Again, the tinker figure is entwined in craft, trickery and subterfuge, an anti-authoritarian dealing with life in his own time and 'in his own strange way'.

But tinkerers (note the 'er' suffix) of this era – those with more material resources than the travelling tinkers (no suffix) – collectively developed many of the everyday technologies that have become central fixtures in our homes and contemporary lives. In his contribution to the anthology *Technological Choices* (2002), French engineer Jean-François Quilici-Pacaud describes a man called G Spratt, a Depression-era figure who 'has been effaced from the history of early aviation', but whose tinkering was integral to engineering the aircraft stability we rely on today. Likewise, American social historians have described photographic records from Kansas during the Depression era that show how farmer Bill Ott and his daughter Lizzie tinkered

Howard Wright and his daughter Evelyn with Wright's tinker wagon, Narrandera, NSW, 1953. Photograph: Jeff Carter, courtesy Jeff Carter Estate and the National Museum of Australia.

Far left: Detective John Christie disguised as a travelling tinker. Photograph: John Mitchell, 1845-1927. Courtesy State Library Victoria, Herald and Weekly Times portrait collection.

Left: Tinker near Hamilton, Victoria, on his travels with his dog and her pups. Undated but print from negative 1947–1955. Photo: Ursula Powys-Lybbe. Courtesy State Library Victoria, The Touring Camera in Australia Collection.

with their car (a Ford Model T) and invented a motorised washing machine. Removing the car's rear tyre and adding a drive belt, they built a homemade ethanol- or gasoline-powered machine that washed clothes. This was around the time engineer Carl Miele and his business partner Reinhard Zinkann, over in Germany, had morphed butter-churns into clothes-washers, and became the first to commercialise mechanical washing-machines.

We may have heard of the Wright Brothers and Miele (whose name is now a multinational brand), but few of us know of the Otts or Spratt, who remain unsung among the everyday people, male and female, who adapted existing technologies in incremental and make-shift ways that became commonplace. Our ignorance of everyday tinkerers, writes historian W Patrick McCray, is attributable to the 'Great White Man narrative of innovation' that ignores 'the critical role that anonymous, unrecognised people … play in the incrementalism that is the real stuff of technological change. Most of the time, innovators don't move fast and break things.' Policy and funding tend to conflate the idea of innovation with entrepreneurship, and most histories perpetuate the myth that great innovation came from the hands and minds of heroic lone white men (from Thomas Edison to Steve Jobs).

But as Bruce Pascoe documents in his powerful book *Dark Emu* (2014), many forms of Indigenous Australians' ingenuity – complex aquaculture and agriculture systems, engineered structures, intricate textiles and handcrafts, hybrid cultivars, inventive traps, netting, preserving and architecture – were for centuries ignored by colonists and settler-descendants, if not destroyed outright. Even when examples of First Australians' ingenuity were documented by early anthropologists and explorers, they speculated that earlier European settlers – not Indigenous 'savages' – must have invented or engineered them. Some scholars now maintain that First Australians – before

1940s war-effort campaigns that promoted thrift and home-based production as patriotic duty.

the Egyptians – invented bread, and researchers are only now attempting to revive some age-old Australian techniques, systems and cultivars in order to return a sustainable stasis to Australian soils and farming practices.

'The driving forces of innovation are not mythic isolated geniuses, almost always represented as men,' writes McCray in his *Aeon* essay, 'It's not all lightbulbs'. 'This view of innovation – narrow and shallow – casts a long shadow, one that obscures the broad and deep currents that actually drive technological innovation and shape its impact on society.'

By the time power tools made their way into the domestic sphere in the mid-1950s, the tinker figure was revived, romanticised and bohemianised in Left Bank Parisian culture. In Australia, during both world wars, ideas of tinkering – as thrifty, resourceful and inventive DIY practices – were endorsed in nationalist campaigns. Wartime Britain's 'make do and mend' war-effort campaign (ending in 1946) had cultural impact here, to the extent that home-based self-sufficiency and resourcefulness were promoted as patriotic duty. The 'victory garden' campaigns during both world wars promoted the Australian home as a site of self-sufficient and national duty. During the First World War, *The Motorists' Almanac* advised domestic car-owners that 'Much pleasure and a great saving of energy can be derived from undertaking your tinkering yourself … odd half hours expended in these humble jobs promote the brotherhood of man and the machine.'

But this collaboration wasn't just of 'man and the machine'; during the war period women, too, became hardware tinkerers. Historian Kathleen Franz describes instances when:

> Women drivers tinkered with the car and, by extension, with their gender roles. Stories of motor travel written by and for women in the period between 1910 and 1920 revised the dominant discourse of ingenuity to include women as technological heroines … if short-lived, role models …

During the Second World War, Australian prime minister John Curtin's 'all-in' war effort campaign, rationing clothing, food and fuel, saw thrift and home-based manufacture promoted as a moral virtue – even to wealthy households. Unorthodox inventive practices that in peacetime were not tolerated were readily accepted in wartime in allied countries, as the desire for victory eclipsed other cultural agendas. And until the postwar period, motor cars served tinkerers' many purposes – it wasn't uncommon for them to be jacked up in paddocks and adapted to pump water, grind grain, saw wood or churn butter.

Initially, the Ford Motor Company endorsed this type of tinkering – historian Ronald Kline documents many instances when the company itself promoted hacking and adaptation. In one of the earliest published photographs of a car being used to generate power, Henry Ford is sawing wood with a new Model A Ford, with a crankshift inserted into its side, and a long shaft attached to this that runs to a pulley. After the release of other models, *The Ford Times* also reported how each model was suited to different tinkering, and that 'with a little ingenuity the engine can be made to run the cream separator, saw the wood, or pull a trailor'. A whole industry emerged of elaborate conversion kits, designed to overcome the undue strain on the differential gear that jacking up one side caused, and taking power directly from the axle or crankshaft. At least 22 start-up companies made these kits.

But in postwar years, these industries rapidly died out as car manufacturers started discouraging repair and adaptation. The automotive industry, formerly encouraging home repair, 'characterized tinkering as useless and inefficient,' according to historian Kathleen Franz. And Ford issued missives to dealers forbidding them to encourage 'makeshifts', threatening withdrawal of their dealership licences if they continued to sell conversion kits. This shift in manufacturers' values and power-relations was reflected in magazines. While the influential *Popular Science* featured home-based construction and electronics projects (including car and appliance repair) into the early 1950s, the war-free times would see these replaced by articles about Big Science.

At a time when 'handmade' was associated with 'poverty', the promotion of intimate material literacy with household technologies was replaced with that of a distancing awe. Productive home technologies such as sewing machines, domestic shed-tools and increasingly 'professional' food appliances and recording devices had rapidly increased in demand since industrialisation, and until the postwar period these carried instruction for repair and modification. But towards the 1960s and 1970s home-appliances became 'tamper-proof' – resistant to home-repair, a continued predicament against which contemporary tinkerers protest (see Makers' Bill of Rights).

Accordingly, The Terrible Tinkerer, a gadgeteering geezer who meddles with contraptions with which to blackmail decent law-abiding folk, started appearing in *Spiderman* and popular comic books. Dr Seuss introduced Sylvester McMonkey McBean, the underhand travelling 'Fix-it-up Chappie' who duped the hapless Sneetches with his inventions. Again, the tinkerer became a deviant travelling figure in popular culture, and subsequent popular perceptions of tinkering as 'keeping busy in a useless way' could be understood as a secular inversion of 'idle hands are the devil's tools'.

Dr Seuss's travelling 'Fix-It-Up-Chappie', who appeared in *The Sneetches and Other Stories* (1961).

Right: The Terrible Tinkerer, aka Phineas Mason, who first appeared in 1963. (Open source: Wikia Marvel Database).

Far right: A contemporary version of The Terrible Tinkerer. (Open source: Marvel Wikia database)

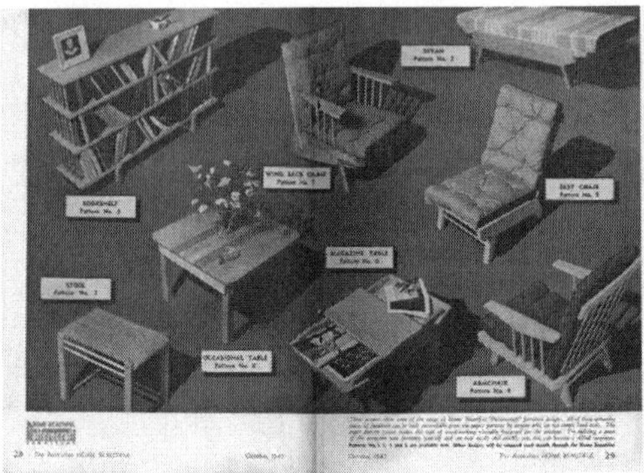

Cover of October 1947 *Australian Home Beautiful* magazine and Patterncraft Furniture featured in October 1947 Australian Home Beautiful magazine, courtesy Pacific Magazines.

In postwar Australia, DIY furniture became commonplace, as patterns became available by mail-order through magazines including *Australian Home Beautiful* (pictured). Three major brands of furniture kits emerged, with cut pieces ready to assemble by laypeople 'with limited skills or without access to timber,' according to design historian Nanette Carter. Such kits 'helped Australians furnish their modest homes with objects they had made themselves, avoiding debt.' Many Australians also built their own houses, a trend that 'escalated in the 1950s and reached a peak of forty per cent in 1954.'

When tinker-wagon owner Howard Wright died in 1969, 'the tinker's trade died with him,' according to the *Sydney Morning Herald*. But late-1960s Australia saw a revived spirit of tinkering in the ad-hoc design movement, associated with counter-cultural rejection of consumerist values. At this time, New York and Parisian salons and arts institutions were celebrating *assemblage* and *bricolage* respectively, emphasising materials that were 'discarded or purloined'. To some extent, this counter-cultural ethic was incorporated into capital expansion with the 1970s development of specialist DIY

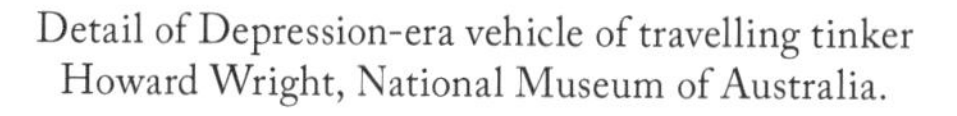

Detail of Depression-era vehicle of travelling tinker Howard Wright, National Museum of Australia.

Boush the Tinker.
Source: Defence of the Ancients, www.playdota.com.

retail stores targeting amateur customers. Prices for entry-level tools dropped dramatically, many hitherto professional tools started being redesigned specifically for the home-based DIYer, and alternative materials (fibreboard, builders' glue, PVC) became workable on a domestic scale, transforming and expanding what the domestic lay-person was able to undertake. More recent sources of normalisation of DIY projects have been makeover programs such as *Better Homes and Gardens*, *DIY Rescue*, *Gardening Australia*, *Backyard Blitz*, *Burke's Backyard*, *Room for Improvement* and *The Block*, and entrepreneurial ones such as *The New Inventors* and *Hot Property*, as well as an explosion of craft and DIY social media. As these programs' popularity peaked, so too did home-renovation in Australia.

In *Renovation Nation* (2008), cultural studies scholar Fiona Allon documents a citizenship ceremony conducted by a federal minister for multicultural affairs in a suburban Bunnings hardware store, and she speculates that 'the most obvious next step would be to hold citizen tests in Bunnings too: What is a router? What do you do with a spirit level? What is the difference between a tenon saw and a dovetail saw? An Allen key and a shifting spanner?' In Allon's speculation, to answer such questions correctly might demonstrate one's suitability for Australian citizenship.

Yet in such industrialised economies as Australia, tinkering's basis in raw necessity and survival have eased, at least for the home-owning classes. Still, survival and revolutionary narratives dominate here and abroad in maker movement discourse. To Mark Thomson, tinkerers are best-equipped to tackle the effects of climate-change; to Seely-Brown, tinkering is now 'critical in a world of constant change'. As I completed my research, *Portland Magazine* editor Brian Doyle depicted tinkering as a claim to humility and modesty. In 'A Tinkering Kind of Guy: The Absence of Arrogance', he profiled Donald Shiley, whose heart-valve invention has saved the lives of half a million people. Doyle reported: 'he [Shiley] says no, he is not great and not heroic, he is just a guy … a tinkering kind of guy, he actually says this, *a tinkering kind of guy*'.

To this day, 'tinker' continues to have itinerant (and other outlaw) associations in various sections of popular culture. Boush the Tinker (pictured), for example, is a resilient, resourceful, enterprising and deviant hero in *Defense of the Ancients (DoTA)*, a popular real-time role-playing game. His 'Boots of Travel' give him 'global mobility', and his 'race is known for its intelligence, its cunning, and its prickly relationship with magic. As a matter of pride, they survive by their wits.'

As much as tinkering may have reached its cultural moment, some of its proponents regard it as an atavistic tradition whose provenance reaches back to the prehensile thumb that first shaped a stone or crafted a pot – and, as such, essential to what it is to be human. 'The pure, focussed bliss of tinkering isn't just for artists or engineers,' write Mike Petrick and Karen Wilkinson in *The Art of Tinkering: Meet 150+ Makers Working at the Intersection of Art and Science* (2014). 'It's a way we can all be more human.' Tinkering is routinely essentialised and has historically been understood in evolutionary terms by science scholars. It has also been seen as a way to *extend* humanity. To *Make*'s commissioned science editor, Curt Gabrielson:

> Throughout history, people have had to go great lengths to get permission or get around laws and tinker with dead people. To the extent that they were able to do this tinkering, real knowledge about body systems slowly amassed. While you might not be able to get a body to tinker with, there's another way to tinker with the way bodies work: build a model of the body's parts.

A very tinkerly solution, but more telling is that Gabrielson, writing in 2014, addressed an audience of insiders thoroughly familiar with the concept of tinkering. This is in stark contrast with broader general discourse, in which *tinker* refuses to cease its semantic shifts. Even a decade ago, it is unlikely that the statements quoted above would carry the meanings they do now, as cultural changes have seen *tinker* disrupted, revised and subverted. We have reached a moment in history where *tinkering* is now a contronym. To insiders, tinkering is positive and productive; to outsiders, it remains meddlesome and unproductive, an 'attempt to repair or improve something in a casual or desultory way, often to no useful effect,' according to the Webster dictionary.

WIND US UP AND SET US LOOSE

A CODIFIED SET OF THE BUILDER'S, CRAFTER'S, MAKER'S

NO NEED SHALL STAND UNADDRESSED. There is a way to make things better.

NO CRAFT SHALL PASS UNAPPRECIATED. It is always worthwhile to pause and intake a thing of beauty. Plus, it may spark a concordant idea of one's own.

THE EFFORT ITSELF IS THE REWARD. Yes, I did spend three days on a device to save me ten minutes. You are missing the point of the process.

A FAILURE IS SIMPLY AN EXPERIMENT. Most efforts will fail. Rarely should the same failure repeat itself.

MISCELLANEOUS PARTS ARE LIFE. I'm keeping that because I will need it someday. Last time I threw something out, I needed it the very next day.

PROBLEMS ARE DESIGNED TO BE SOLVED. Spring into action. Make it happen. Hannibal was right: I *love* it when a plan comes together.

PROJECTS ARE STACKABLE. It's not that I'm starting something new before finishing something old—I'm *nesting* the new project *inside* the old.

IF IT HAS SCREWS, THEY SHALL BE TURNED. If it has bolts, they shall be loosened. If it has rivets, they may get along for a while without being pried open, but that probably won't last.

IF IT IS BROKEN, IT IS FAIR GAME. Maybe I can fix it. Maybe I can use some of its parts for something else. Maybe I just want clearance to be more rough than usual when opening up the case.

THE INSIDES OF THINGS ARE BEAUTIFUL. Let's see what they look like.

TAKE IT APART

wondermark.com

makezine.com

THE MAKER'S BILL OF RIGHTS

■ Meaningful and specific parts lists shall be included. ■ Cases shall be easy to open. ■ Batteries shall be replaceable. ■ Special tools are allowed only for darn good reasons. ■ Profiting by selling expensive special tools is wrong, and not making special tools available is even worse. ■ Torx is OK; tamperproof is rarely OK. ■ Components, not entire subassemblies, shall be replaceable. ■ Consumables, like fuses and filters, shall be easy to access. ■ Circuit boards shall be commented. ■ Power from USB is good; power from proprietary power adapters is bad. ■ Standard connectors shall have pinouts defined. ■ If it snaps shut, it shall snap open. ■ Screws better than glues. ■ Docs and drivers shall have permalinks and shall reside for all perpetuity at archive.org. ■ Ease of repair shall be a design ideal, not an afterthought. ■ Metric or standard, not both. ■ Schematics shall be included.

Make:

MENDRS MANIFESTO

MAKE MENDING VISIBLE

MAKE MENDING VALUABLE

MAKE MENDING SOCIABLE

MAKE MENDING POLITICAL

MAKE MENDING DESIRABLE

MAKE MENDING VITAL

MAKE THINGS MENDABLE

MAKE MORE MENDERS

MAKE LESS AND LESS

MEND MORE AND MORE

REPAIR MANIFESTO

WE HOLD THESE TRUTHS TO BE SELF-EVIDENT

IF YOU CAN'T FIX IT, YOU DON'T OWN IT.

REPAIR IS BETTER THAN RECYCLING

Making our things last longer is both more efficient and more cost-effective than mining them for raw materials.

REPAIR SAVES YOU MONEY

Fixing things is often free, and usually cheaper than replacing them. Doing the repair yourself saves you money.

REPAIR TEACHES ENGINEERING

The best way to find out how something works is to take it apart.

REPAIR SAVES THE PLANET

Earth has limited resources. Eventually we will run out. The best way to be efficient is to reuse what we already have.

REPAIR CONNECTS PEOPLE AND THINGS | REPAIR IS WAR ON ENTROPY | REPAIR IS SUSTAINABLE

WE HAVE THE RIGHT: TO DEVICES THAT CAN BE OPENED | TO CHOOSE OUR OWN REPAIR TECHNICIAN | TO NON-PROPRIETARY FASTENERS | TO REPAIR DOCUMENTATION FOR EVERYTHING | TO REMOVE 'DO NOT REMOVE' STICKERS | TO REPLACE ANY & ALL CONSUMABLES OURSELVES | TO TROUBLESHOOTING INSTRUCTIONS & FLOWCHARTS | TO REPAIR THINGS IN THE PRIVACY OF OUR OWN HOMES | TO ERROR CODES & WIRING DIAGRAMS | TO AVAILABLE, REASONABLY-PRICED SERVICE PARTS

REPAIR IS INDEPENDENCE SAVES MONEY & RESOURCES | REQUIRES CREATIVITY | MAKES CONSUMERS INTO CONTRIBUTORS | INSPIRES PRIDE IN OWNERSHIP

IFIXIT JOIN THE REVOLUTION WITH IFIXIT.COM

Left page: Wondermark's A Codified Set of the Builder's, Crafter's, Maker's Rules, by David Malki ! available at wondermark.com.

Top left: Mister Jalopy's A Maker's Bill of Rights.

Above: iFixit's Repair Manifesto.

Bottom left: Lancaster University's MEND*RS research symposium manifesto.
(Source: MEND*RS).

As late as 2015, the few subeditors who remained on the Fairfax Media payroll were placing quotation-marks around the term *maker movement*, as if it were a novel and unfamiliar concept. But the term had been generated from Silicon Valley for at least a decade. And it had such strong cultural currency that the year before, in 2014, it had already been thoroughly satirised in sections of popular culture (as I'll discuss).

Tinkering is a defining feature of the maker movement. The maker movement has been characterised as many things, but is largely understood as a rekindled DIY movement, a collective desire to make, adapt and repair products rather than buy them off-the-shelf. It's also portrayed as a tech-enabled democratising and revolutionary force. In fiction and non-fiction alike, maker movement literature is loaded with cheerleading heroicism and revolutionary rhetoric, promoting material DIY as the pathway to individual happiness, entrepreneurial success and cultural reform. (More on this in Chapter 8.) In his novel *Makers* (2009), open-copyright activist and bestselling author Cory Doctorow depicts off-the-grid networks of micro-financed DIY survivalists (3D printing entrepreneurs, technology tinkerers, garage repurposers) who thrive using their tinkerly wits amid the collapse of standardised and corporatised mass-production. And in his non-fiction *Makers: The New Industrial Revolution* (2012) Chris Anderson portrays a similar vision – a maker movement future of 'ever-accelerating entrepreneurship and innovation with ever-dropping barriers to entry'. Anderson heralds a 'Third Industrial Revolution' which is 'extending manufacturing to a hugely expanded population of producers – the existing manufacturers plus a lot of regular folk who are becoming entrepreneurs'.

There's some scholarly quarrelling about whether the maker movement is actually a social movement, a faddish brand, or just a marketing meme dreamed up by *Make* magazine founder Dale

Doherty to attract now-millions to world Maker Faires. But its existence as a revolutionary force is taken as gospel by many scholarly and media reports, and by 2014 the maker movement's utopian promises had already been satirised on a grand scale. In what could be read as a traditional manufacturing backlash against the movement's reformist promise, Honda launched a spoof DIY-car advertisement on 1 April, in a style derivative of the hipster-bashing TV series *Portlandia*. It was viewed by several million. In it, self-described tinkerers Caleb and Vanessa brew their own kambucha in their loft. She collects handmade sheep yarns 'from España'; he is 'very excited' about his collection of vintage blown-out amplifier tubes and fuses, from which 'I'm planning to make a sort of, like a, collage of my mother's face'. Honda's DIY build-your-own car arrives 'just like our organic produce, only a lot more boxes'. By producing these effete caricatures, Honda, according to *Slate* magazine, was 'taking aim at the absurdity of the overly self-serious DIY fad'.

Caleb: I guess we are what you would call …

Vanessa and Caleb together: Tinkerers.

Vanessa: We consider ourselves part of the whole handcrafted movement.

None of this book's tinkerers could be fairly described as self-serious or faddish. A couple believed the advertisement was engaging in a strawman or stereotyping backlash. Mark Thomson told me: 'I felt a tiny bit sick when the first scene in that ad was "we think of ourselves as … tinkerers!" … There's a vast gap between the maker world and the manufacturer world.' This distinction has largely been lost in the hyperbolic claims of maker movement discourse that still dominates media reports. Indeed, Honda was satirising a discursive reality rather than the realities underway in people's backyards, kitchens, paddocks, craft rooms and sheds.

Against the maker-movement hype, the people profiled in this book frequently explain what tinkering is *not*. Mark Thomson, for example, said tinkering 'is not goal-directed nor are there defined outcomes. There are no key performance indicators for tinkering … [it is] not a product or a saleable process.' Outside maker movement literature, tinkering is currently recognised as a transformative creative practice, an important method of knowledge-production, a method of self-sufficiency, and the genesis of a vibrant informal economy. It's lauded as an egalitarian vocation that has reached its historical moment. 'I'm a tinkerer,' Rolf Hut told TED audiences in 2011, explaining that he likes 'to take things apart, and see what makes them tick. And then, make them tick harder. Or faster. Or make them do stuff they were never intended to do.' Hut, a trained scientist and senior academic, doesn't list these as his vocation – his bio-note states simply: 'Rolf Hut is a tinkerer.' He continues: 'In the old days, you needed a lot of money, and a lot of time, to be a tinkerer … Today everybody can be a tinkerer.'

Tinkering's moment is now, echoes Creative Commons proponent Marleen Stikker, who in *Open Design Now* (2011) describes tinkerers as 'the pioneers of our time' who 'are not taking the world at face value, as a given from outside; rather, they see the world as something you can pry open, something you can tinker with'.

And Steve Daniels, editor of *Makeshift*, writes: 'Do not underestimate this [tinkerer] demographic. The informal economy of unregistered and unprotected enterprises accounts for over three quarters of employment across Asia, Africa, and Latin America. These maker-entrepreneurs are resilient, flexible, and immensely creative.' In post-industrial economies, too, the word 'tinker' is enjoying a renaissance. Data searches show us that the terms *tinker*, *tinkering* and *tinkerer* are enjoying unprecedented currency in education and social media. As I completed my research, the Swinburne Tinkerers' Guild formed at my university. Tinkering education programs are emerging in the US and in Australia; state institutions are launching tinkering programs; fixers' collectives and hackerspaces are surfacing; tinkering conferences are being organised; sites like *tinkercad* and Microsoft's *Tinker* are established; and a cursory read of *Make*, *Makeshift*, *Popular Science*, *Boing Boing*, *iFixit*, *Toolsparesonline.com*, *Gizmag*, *Wired*, *RENEW* and *IEE Spectrum*, or the broader-market *Smith* and *Manspace*, reveal that many people now identify as tinkerers.

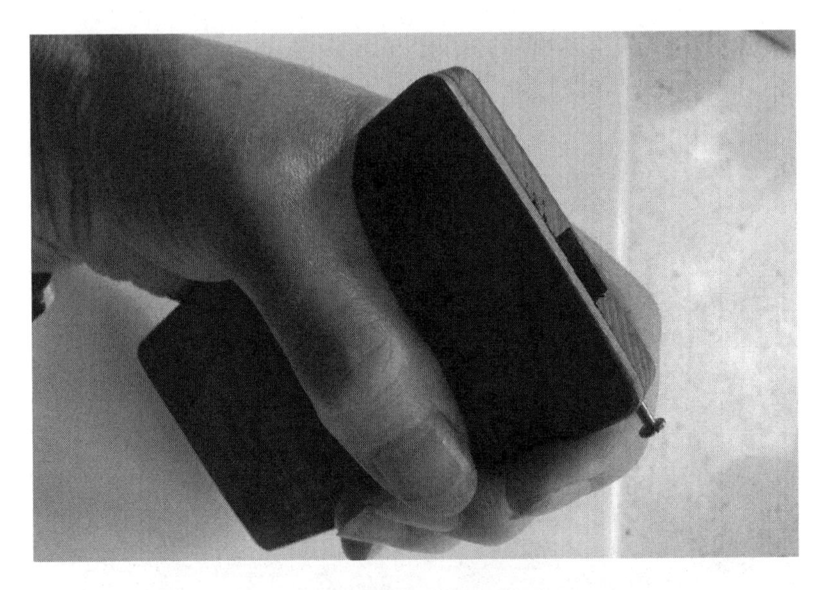

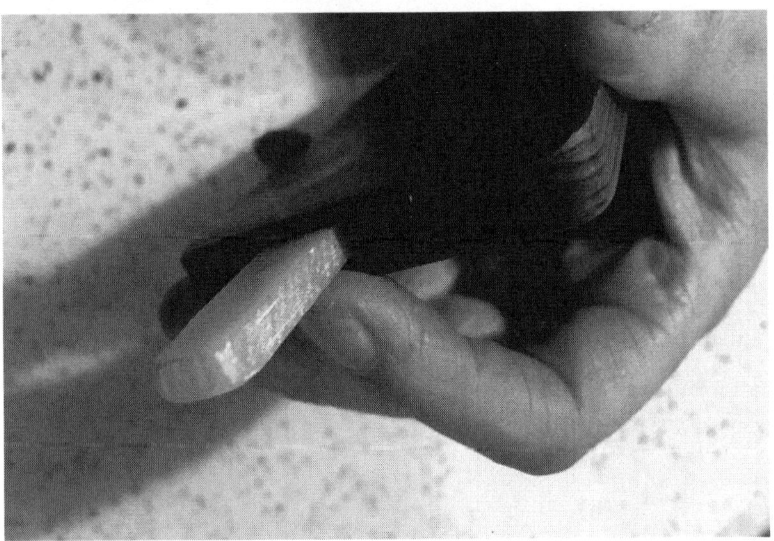

Chris Block's Eye-Slate with its nail fastener; and the Eye-Slate with a split of engineer's chalk emerging from its orthotope slot.

3

Magic

I visited Chris Block in the pastoral slopes of South Australia's Carey Gully, where he built his north-facing home atop two large round concrete water-tanks. (He built those, too.) A silver-haired man, Block seemed too soft-spoken for his Hard Yakka overalls. In his unfettered enthusiasm for tinkering, Block appeared every bit as spellbound by materiality as the next tinkerer – but he spoke with an exceptionally reverential marvel of the workings, forms and possibilities of materials.

The son of a builder and himself a builder of boats, homes and furniture, Block had invented many things. Among these were agricultural systems, commercialised and patented chair designs, motorised bicycles, and an unorthodox method of terrazzo. I planned to question him on what to me seemed his simplest invention: the iSlate. (Later, in an email, Block revised the spelling as 'eye-slate'.) With no moving parts, crafted out of slate, wood, stone and steel, Block's iSlate was immediately recognisable as the ubiquitous form of an iPhone. He would later tell me:

> One of the things that I personally enjoy is smooth objects. Mr Apple got it right, and I copied that. The slate feels good, the timber feels good, but the feel tells you this is a piece of quality. It's not a piece of shit that'll go straight into landfill. It has its own story. The shape itself has magic.

I first encountered the iSlate – whose name is now trademarked by Apple Inc. for an altogether different gadget – on a plinth within a travelling exhibition of bushpunk contraptions Block had made with his tinkerer-comrade Mark Thomson. It first appeared to be a low-tech data storage tool. Fronted with a slate veneer fused to a cedar body, it contained a slot-pocket in the wood to accommodate a split of engineer's chalk. A removable flat-head nail wedged across the slot's opening secured the chalk. Viewed in the context of a gallery, it could be understood as art: a witty take on time and technology.

But Block didn't describe his objects as 'art', even though from time to time he exhibits in public galleries. He described his objects as 'things', 'tinkering' or 'projects'. Even those he described as 'machines' weren't necessarily efficient or useful. With Thomson he'd made, among other oddities, a Random Excuse Generator. Ostensibly running on pedal-power generated to triangular cogs, this machine featured a bakelite platitude accumulator dial, a blame shifter, verb gas flow, and letterballs that blew like lottery balls in a basket. When powered from its pedals, it displayed random political excuses across a screen. Block and Thomson lugged this Random Excuse Generator onto a trailer and hauled it to a farmers' field day, where rural engineers were showing off their restored farm machinery. It invited mixed reception: many laughs, but some tut-tutting. Thomson and Block told me they were accused of having too much time on their hands. One visitor offered a variation of 'idle hands are the devil's tools'. Thomson told me later:

> Chris [Block] and I just thought it was funny. These [the naysayers] were people who just spent three-and-a-half years restoring a rusty steam engine that pumped water into a pipe. Nothing spectacularly useful. Their problem with it was that they could see that we clearly knew how to make something work.

When I saw it exhibited at the Murray Bridge Regional Gallery, though, the Random Excuse Generator invited certain interpretations. If not studio art or consumer technology, then maybe folk art or folk technology. But tinkering exposes these categories – art as

Mark Thomson and Chris Block's Random Excuse Generator.

distinct from technology – as the modish contrivances they are. Our ancestors would have regarded these distinctions as absurd. Recent histories describe Roman ideas of *ars* and Greek ideas of *techne* as more-or-less interchangeable in classical antiquity. *Techne* has often been translated as 'to make', and, according to historian Larry Shiner, both these terms 'embraced things as diverse as carpentry and poetry, shoemaking and medicine, sculpture and horsebreaking. In fact, *techne* and *ars* referred less to a class of objects than to the human ability to make and perform.' *Poein*, the verb for poetising, also meant 'to make'; *poiesis*, meaning 'composition' or 'making', was the work of the artisan or craftsman.

For a few centuries before the Enlightenment, *techne* and *ars* were generally grouped to a core of seven making-practices: 'most frequently medicine, agriculture, mechanics, navigation and gymnastics', with poetry being a subcategory, variously placed under differing taxonomies (at one stage under 'physics'; at another stage under 'rhetoric'). In classical antiquity, there was no discrete term to describe visual arts. Cultural scholar Camilla Nelson notes that the idea of 'creativity' is also a very recent one – there was no word equivalent in meaning to 'create' in ancient Greece. Nelson writes that '*techne* could be – and was – applied to all forms of human endeavour from verse making to shipbuilding and bricklaying'. These broad understandings aren't limited to Graeco-Roman origin – in Mandarin there is still *Yi*, variants of which can be translated as skill, art, wits, talent and craft; in Hindi there is *kala*, which can be used for science, art, skill, ability, cunning and magic; in Vietnamese there is *nghệ thuật*, which can describe science, art, technicality and technique. The Portuguese *arte* can describe science, knowledge and artistry.

Tinkering, too, invites us to devolve and broaden our ideas of 'practice' and 'making' in trans-historic and universal ways. Many recent histories have described how English-speaking cultures' ideas separating art from technology (or art from science; or art from engineering) didn't fully emerge until the mid-18th century, alongside secular and industrial institutions (in which we also classified many Indigenous artefacts as 'art' within our museums and galleries). Historians of European art have noted that even as late as the Middle-Ages, art wasn't understood as aesthetic or conceptual *tout court*, but as functional and scientific, like medicine and cartography. This reminds us that, as anthropologist David Graeber puts it: 'what we assume to be immutable has been, in other times and places, arranged quite differently, and therefore, that human possibilities are in almost every way greater than we ordinarily imagine'. Contemporary ideas of a set of creative pursuits as distinct from engineering or scientific ones aren't at all useful when we consider tinkering. But the expansive, magic possibilities of *techne* and *ars* are.

I visited Block in pursuit of the iSlate's story, but by the time he'd set down a mug of coffee the story had digressed into that of another project. 'Because of that exhibition,' Block explained, 'the people at [South Australia's] Bay Discovery Centre [Museum] said, "Ooh, you guys can do stuff". They could see that we knew how to make things work.' The museum administrators offered the two tinkerers a $20,000 commission to make a *Titanic* exhibit. The exhibit was an interactive display that would run on clockwork internally but also invite public operation via a mechanised ship-wheel. 'It's a time-display,' Block explained. 'Of what happened during the twelve hours at the sinking of the *Titanic*.'

I asked Block if he'd ever made such a thing. No, he'd never made anything like it. But he had no doubt he could pull it off. Tinkering is a practice of domain-shifting your prior knowledge. 'When you've put machines together in the past, you tend to be able to work things out.' Once he'd worked out how to dissemble and fix his first engine (a Volkswagen) many years ago, this gave him confidence to work out *any* machine:

> I put a Volkswagen together from the crankshaft up. Most mechanics hate them. They don't like them because they were different. Having done that, machines don't frighten me. I know I can do it. One of the things I did when I was first separated from my first wife is I sold the house. I bought a boat, the boat wasn't as good as I thought, so I had to learn how to repair it. If you live on a boat, you have to be able to repair things, your survival depends on it. I learned to be a boat-builder, and then worked as a boat builder …

'People are scared of failure,' he told me. 'I make mistakes all the time, that's how you learn about things and get more experienced. You have to *think*, and you have to be patient.'

'But that's just it,' I interrupted. 'Patience takes committed time. Most people probably don't have the time to work things out.' As the following chapters show, tinkerers tend to organise their lives around tinkering, rather than fitting into externally-imposed work and leisure regimes. Block told me he feels sorry for people who aren't masters of their own time. He said:

> I always have this argument with my wife, Julia [a film producer]; she asks why I want to do these things [like the *Titanic* commission], and I explain to her I want the money so I can *buy back my life*. To buy back my life, you know, so all these other bastards don't have control of my time.

Like other protagonists, Block's tinkering was his central life focus. The *Titanic* project was to a large extent self-directed, but it was in the end a commission, not strictly autonomous. So did this project count as tinkering? It turns out that's the wrong question. Tinkering isn't a discrete activity but a way of life, involving trade-offs between paid and unpaid, external and autonomous work regimes. Simply by devoting your own time, Block said, you build confidence to tinker with things that experts say can't be tackled, because by doing so 'things become demystified'. And this has economic impact, freeing up money for other projects. Over Christmas, for example:

> I had to take the gear box out of the Land Rover and pull it to pieces because the local mechanic said he couldn't do it. So I did it. I'd never pulled a gearbox apart [but] … it's not that hard.

Having worked out the task himself, Block didn't need to pay the mechanic, and fixing his own gearbox gave him knowledge for the next project (and the next saving). But, to Block, tinkering wasn't simply about thrift and self-sufficient pride. Nor were his motivations merely to understand machines conceptually. He was vested in the deep pleasure of understanding them tacitly, with all senses:

> Now, when I change gears [in the Land Rover], I visualise, I can *see* and *feel* those cogs engaging. When it goes click-click, click-click, you know *exactly* what's going on. When you put something together, you *know*: every click you *understand*, you *know* what's going on.

Descriptions of haptic feedback are routine in tinkerers' stories. It's through these bodily entanglements and enchantments with their materials that tinkerers *become* their projects, in ways that can be transcendent. Sociologist Ben Highmore views the body as a 'nexus of finely interlaced force fields', and the celebrated science sociologist Bruno Latour views our bodies not as discrete material entities, but as sensory mediums through which 'we learn to register and become sensitive to what the world is made of'. Other scholars describe the synergy between the body and materiality as if it extended to boundless, almost panpsychic breadth. In her 2008 book *My Stroke of Insight*, American neuroanatomist Jill Bolte Taylor explained that when she suffered a stroke, she could no longer discern the difference between the molecules and atoms that constituted her arm and those that constituted her bathroom wall. Once recovered, Bolte Taylor didn't view her 'other' perception as delusional or debilitated, but as an important way of experiencing the world, and part of our healthy perceptive possibility. Neurological studies documented in books such as Frank Wilson's *The Hand* (1999) and David J Linden's *Touch* (2016) affirm that the interactions of our bodies with material things shape our brains (literally, architecturally, in their molecular structure and circuitry) – just as the brain directs the body (and its muscle-memories) to shape materiality. Technology historians have pointed out that our collective prose became less fluid and more staccato with the advent of the typewriter; neuroscientists have documented the ways taxi drivers' hippocampi (brain memory centres) ballooned with detailed spatial memory of London streetscapes, only to deflate again when GPS technology was introduced; it's accepted in musicology that a recursive feedback loop is generated between mind, hand and instrument. From his prolific studies of material culture, British anthropologist Daniel Miller concludes that 'objects create subjects much more than the other way around'. If you tinker, your body isn't simply a tool with which to impose your conceptual will onto materials: your body and your humanity meld with materials and *become* the tinkered project.

During one visit to Block, he lapsed into a rhapsodic and tangential story about the behind-the-scenes happenings inside various machinery: machinery that could conceivably power the *Titanic* museum display. And also the *actual Titanic*. It wasn't easy for me to discern which was which, but the story's trajectories were fable-like. 'The trick,' Block said halfway in, 'is the sequence of events'. The turning of a cog, the response of a piston, the pivot of a lever, the timing of a gear – these are all *events* in the tinkerer's story. A working machine is a festival of such events. Each event involves various quests, problems and arcs: the angle and pitch of propeller blades for optimum velocity; the problem of water displacement when using propellers not curved on both sides; the function of pulley-blocks and capstan winches driven by 12-metre levers; the problem of an axle levering anti-clockwise without affecting the direction of other components; the ways electrical impulses behave like water molecules; the uses of a ratchet drive.

In sum, to make the museum display's internal moving parts work, Block needed to source two very specific *two-way* gearbox cogs, on which to mount a bike-chain. Nowadays, these very specific two-way cogs are not at all easy to come by, and here Block gave an example of a recurring tinkerer's trope: the magic incident of the serendipitous finding. Tinkerers are sometimes hoarders, or when they can't source something online they often rely on hard rubbish or on hoarder-suppliers. Happily for the *Titanic* project, Block's neighbour, like Gogol's Plyushkin, populates his paddocks with all manner of accumulated junk awaiting revival or decay. In Nikolai Gogol's 1842 novel *Dead Souls*, Plyushkin is a landowner-hoarder, and 'Plyushkin Syndrome' became a Russian byword for the psychology of such a character. It's difficult to fathom why such an expression doesn't feature prominently in the Australian vernacular or literary tradition. Countless farms I've visited (including broadacre and livestock farms in my own family) have paddocks populated with scrap metal;

Two-way cogs, salvaged out of agitator mechanisms of washing machines.

countless suburban backyards I saw during my research were libraries of organised debris; and hoarding has also become a fashionable topic for academic investigation and television series alike. Many hoarders are befriended by tinkerers, and they serve as rich material resources as well as characters in tinkers' stories.

So Block and his hoarder-neighbour set to work scavenging the paddocks for the rare cog, and after many bum-steers finally found a 1940s washing machine. They rummaged around its innards for the agitator mechanism, and were disappointed to find it was 'badly-rusted. We couldn't use the shaft and bearings.' *But*, he added, 'it had the right cog!' This was a triumph, but the quest didn't end there, because Block needed *two* such improbable cogs, and finding a second seemed ludicrously unlikely. But the two kept faith and pushed on. By this time Block's story had taken on dramatic suspense. After much adventuring, Block and the neighbour had all but given up; but then, 'we took one more step in the dirt, and there was another cog. Just sitting there, in the dirt!' The improbable serendipity story isn't just commonplace, but almost without exception among the tinkerers I studied. Melbourne tinkerer John D'Alton described such serendipity as 'magic'.

A couple of months later, Block described the *Titanic* project's subsequent quest. It needed a very specific, custom-made Geneva gear, a two-wheel mechanism. One wheel is the drive wheel, the other is the driven wheel. This mechanism, commonly used in wind-up watches and film projectors, converts smooth rotation (drive wheel) into intermittent rotation (driven wheel). The number of steps in the rotation is determined by the number of *slots* in the driven wheel. A clock has 12 slots per day for the hour hand and 60 per hour for the minute hand; four slots will produce four 90-degree rotation steps; eight slots will produce eight 45-degree rotation steps, and so on.

For the *Titanic* project, Block needed a driven wheel that contained 12 slots, to give round-the-clock step-by-step window-stories about the *Titanic* sinking. The time, though, could be reversed or accelerated by museum visitors operating the ship-wheel by dint of the salvaged two-way gear-cogs. Having researched this project, Block realised that it:

> had to have bearings, not just shafts because then it's not going to work. If you have a look inside a clock, it's got shafts and cogs and all sorts of things, but then this one has to be strong enough for the public.

For this level of public use, the Geneva drive needed not simply a pin, but a sealed bearing which provides its own lubrication and can withstand a load. This needed more precision than a salvaged find could offer:

> Using sealed bearings means I could bolt them. I started looking at how I could put things together, I found a guy called Chris Canon. I used to work for his father – he can convert things into pdf files on a computer, and so a laser cutter can read things.

I asked: 'You mean this man has a program that can translate technical drawings into Laser-cut pieces of metal?' 'Exactly,' answered Block:

> I drew the technical drawings, and Chris Canon converted them into computer language – the engineering details of the actual gears he worked out with a computer.

Block inflected 'a computer' in the uneasy generalising way that you used to hear among web neophytes who described the discovery of things on 'the internet' – as if it's something otherworldly, abstract and unknowable, like the luminiferous ether. He told me later that he was still learning online literacy – he hadn't yet dedicated much time to computers, and he didn't yet know how to use CAD – he hand-draws:

> [I had to] design the positions of everything. There's two frameworks in this machine, and four axles, all this depended on Mark's design, and my interpretation of that. I sourced all the bolts and worked out the positioning.

The spectator of the *Titanic* contraption might hear the clank and rattle of its internal workings and perceive the romance and thrill of Victorian machinery – but to Block, these noises tell an altogether different story. When he first put the machine together 'it kept running on, and disengaging'. Block told me:

> He [Chris Canon, the laser-cutter] realised that you could make it engage a lot more simply by making the Geneva gear crescent more convex. We quickly whipped one [a prototype] up with the bandsaw with plywood, just drew it up with the compass. It was a little bit sloppy, so I cut along the line, and just sanded a little bit ... When we had the gears cut, Chris said I have a third of a millimetre tolerance, I thought: *Shit*. The only way to test it is put it all together, we just don't have the time to make mistakes. We were so frightened, we weren't sure of the tolerance of this laser cutter. I've allowed a bit of a tolerance, normally there is adjustment possibility [but for this project] it's got to be right on or not. When it's clonking, gears clattering, it sounds good for the audience. [But] for me, it's oh well, it's not an issue, not a real problem.

Block hears the 'it's clonking, gears clattering' as 'she'll be right' imperfection and improvisation. I suggested that the museum visitors might experience these same sounds as an important part of the machine's magic. He replied:

> Yeah, I think so, and I suppose, with the Geneva drive, it will be magic, it actually *is* magic, because they can't see it, and wouldn't have a clue. There's not magic for me because there's no mystery, but it's magic of quality machines I find wonderful. Things that are repairable, made of beautiful materials that are made beautiful, is magic to me.

But when I asked him if knowing how machines work dampens some of their magic, he replied: 'No! When you know how they work they become even *more* amazing!' For sure, when you Google the components of Block's projects, and Wikipedia offers you animated gifs of Geneva drives, pistons, gears, belts, propellers and crankshafts, beholding how these things work is mesmerising and thrilling. When you take the time to watch intently, you find yourself marvelling at humanity's improbable genius. Then, when you immerse yourself in this marvel while starting your car engine, it does feel like a kind of epic magic. You find you can attune to the workings beneath the bonnet, and experience its rhythms and thrums as if you were witnessing a life-force expressing itself. It can elevate you much in the way good art or a musical crescendo might; you can comprehend Victorian ideals of a 'brotherhood between man and machine'; you can now begin to understand why rev-heads rev, why grease-monkeys grease. And you can feel like a philistine for not having understood before now these unsung people (the Ricky Muirs of our world) and their fine attunement to the arts of the machine.

Tinkerers have a profound camaraderie with their machines. Adrian Matthews, a tinkerer I met in the South Australian town of Macclesfield, was especially attuned to an operatic chorus that

Adrian Matthews with one of his biofuel-converted vehicles.

each of his biofuel-converted vehicles produced. His deep pleasure in their miraculous vitality was palpable, and part of this miracle stemmed from their against-the-odds narratives. A few mechanics had told Matthews that his unorthodox conversions of engines from mineral-diesel to crude used vegetable oil wouldn't work, but Matthews lay awake night after night imagining how they could, and, through a series of tweaks, they did (see Chapter 7). This pleasure and camaraderie with machines can also be ancestral. Philosopher Kevin Kelly offers a theory of the *technium*, a self-generating system that encompasses cultural ideas, social institutions, laws, technological inventions, and – like the German *technik* – 'the grand totality of machines, methods, and engineering processes'. Kelly seeks to supersede our vague ideas of technology (we use that word to describe things as varied as patented products, techniques, instruments, platforms and biotech organisims) so his *technium* encompasses the entire culture of technology that 'behave[s] like a very primitive organism … near biological behaviour'; a self-generating and self-organising culture 'like a very complex organism that follows its own urges'. He, too, reaches back towards early Greek ideas of *techne* as 'something like art, skill, craft or even craftiness. *Ingenuity* may be the closest translation. *Techne* was used to indicate the ability to outwit circumstances.' *Wit*, *wits* and *outwit* are words that frequently appeared in my notebooks when interviewing tinkerers. And so the magic of everyday tinkering is borne of outwitting the present *technium*, and tapping into the ancient one. It lives in mimesis – in plotting yourself in the continuity of experience.

Material objects and the tools that create them are bodily and cultural memories writ concrete. The lemon cake you baked yesterday was borne of centuries of improved citrus and wheat cultivars, a lifetime of honed techniques and collected kitchen tools, a lineage of hybrid layer-hens, and generations of cultural knowledge, trade and technological advances. Baking it can make you feel expansive, part of a continuum. In this way, our tinkered products are a form of memorialisation. Physical expressions of our collective knowledge, they become bearers and extensions of personhood, and like genomes, they carry their pasts within their presence.

Ever since anthropology became an academic discipline, it has described the ancestral magic in the utilitarian practices of everyday life: cooking, quilting, preserving, darning, maintenance – activity that 'not only causes magical materials to be utilizable, but serves to provide them with a ritual character'. Claude Lévi-Strauss, Marcel Mauss and Henri Humbert all described magic as occurring in everyday custom. In *A General Theory of Magic* (1950), Maus wrote:

> Magic is the art of preparing and mixing concoctions, fermentations, dishes. Ingredients are chopped up, pounded, kneaded, diluted with liquids, made into scents, drinks,

> infusions, pastes, cakes, pressed into special shapes, formed into images: they are drunk, eaten, kept as amulets … Magicians prepare images from paste, clay, wax, honey, plaster, metal or papier mâché, from papyrus or parchment, from sand or wood. The magician sculpts, models, paints, draws, embroiders, knits, weaves, engraves. He makes jewellery, marquetry and heaven knows what else.

And in *The Savage Mind* (1966) Lévi-Strauss regarded everyday tinkerers as crafty strategists who live by their wits and construct forms of magic. He described the figure of the *bricoleur* [tinkerer] as an improviser who 'works with his hands and uses devious means … a kind of professional do-it-yourself man'. To the *bricoleur*, 'the rules of the game are always to make do with "whatever is at hand"'. *Bricolage* (tinkering) was understood by Lévi-Strauss as a way of being, a way of understanding and interpreting the world, and also a way of constructing magic systems of connection between people.

When hitting on the possibility of the iSlate, Block initially hoarded old slate roof-tiles. The fine cedar for its body was sourced from hard-rubbish finds: 'The 1920s doors you find are made of cedar,' he said. 'Strip the paint and there it is.' The splits of engineer's chalk, made of soapstone and traditionally used to notate on steel, masonry and timber, he sourced from one of the many engineering suppliers who sell them cheaply in bulk-packs. He bought an expensive diamond-blade for the task of cutting the slate. But the materials refused to submit to the project. He'd found the roof tiles unforgiving to the blade and unpredictable under his power-saw: they split and showed irregularities when cut. So he cadged some offcuts from a bloke at a local slate mine. Trial and error showed Block that the best method was to first secure slate sheets to the cedar with epoxy, and then to rough-cut them to almost iPhone size and gradually refine their dimensions. He could then sand and buff them to replicate the smooth iPhone surface.

But in pursuit of that seductively smooth and curved surface, the project continued to refuse to submit to his will: once the slate was smooth and buffed, the soapstone chalk – valued for its use on rough surfaces – didn't work. Soapstone is harder and less chalky than classroom chalk: to leave its dusty mark, soapstone needs to find purchase within a surface's troughs. After much painstaking experimentation, Block hit upon the exact grade of sanding: one that achieves a convincingly smooth affect but is coarse enough to accommodate the mark-making traction that soapstone wants.

This resolved, there were other twists in the story. How could Block have anticipated the moral battle of the slightly bent 40 x 2.0 mm flathead nail that lay ahead? The quest was to find a way to secure the chalk into the cedar pocket. Various stoppers and valves seemed too fussy and expensive (iSlates could be ordered from Block for around $28; when I suggested a niche market, Block explained that he would limit production, as he didn't want to hear from Apple's IP attorneys). He settled on the elegant simplicity of a removable nail wedged across the slot's entrance. 'The nail,' he explained, 'needs to be bent a little to make it stick in'.

But, without warning, that slightly-bent nail transformed a material thing into a philosophical question, and Block and Thomson quarreled. Some people at the gallery-shop had picked up the iSlates and, habituated by the streamlined logic of contemporary technologies, pulled out the nail and tried to use it – instead of the chalk it secured – to scratch marks into the slate. Thomson considered this a design flaw of the iSlate: it didn't accommodate how people habitually use technology. By contrast, Block argued that the nail was true to the iSlate's *character*: if people covet a non-standardised technology, a bespoke thing, they should rehabituate themselves on the thing's own terms. 'Besides,' he told me, 'the nail is replaceable'. Lose it, and you'll find another in your shed. Like the engineer's chalk and decidedly unlike iPhone parts, it will never be obsolete. Anti-obsolescence is a key ethic of tinkerers, and within its anti-obsolescent narrative lay the iSlate's wit, integrity and magic.

At least, to me. A few months later I phoned Block and he told me that the iSlate had been upgraded. He posted me the new model. The everyday nail was gone: instead it had a pivoting aluminium limb to secure the chalk in its orthotope slot. In making the new iSlate, Block had revised his design philosophy. He'd changed his mind. Clearly, when people tinker with things, those things tinker with people. To me, less had been more – but as the iSlate evolved it shed some of its low-tech wit along with its anti-obsolescence narrative. For me, moving parts sold out the iSlate. I didn't intend to view the thing as art, but, like a jaded critic, I preferred his early work.

I ordered some early-model iSlates for friends, family and my PhD supervisors. As it turned out, the recipients seemed to love them for their wit. All engaged with the cedar, soapstone, slate and steel; some spoke of the *character* of these things. When I suggested that the iSlate might be a downgrade from his current device, my supervisor Julian Thomas (a neophile) answered: 'Not downgrade. Side-grade.' In contrast to Thomas, my (first) principal supervisor Klaus Neumann is the kind of person technology marketers tend to codify in the missionary term 'late adopter'. He caressed the iSlate and later remarked on its properties and events: how the cedar is warm in your palm, but when you turn it over, the slate part is cold. These simple marvels and explanations riffed off the transformative enchantment of everyday materials (a cedar door, a slate tile, a chalk stick, a steel nail) that tinkering enacts. In the tinkerer's *mise-en-scène*, everyday objects are sleeping under a spell of norms, broken by the tinkerer when no-one is looking.

Under the radar, in their everyday sheds and homes, a subterrain of tinkerers work like necromancers, breathing new life into dead and discarded materials; and thus into dead and discarded ideas. In popular culture, these new incarnations are presented as magic. In the British television series *Kevin McCloud's Man Made Home*, repurposed objects (jet-engine becomes a hot-tub; tractor parts and deer skin become a reclining chair) are described by television celebrity Kevin McCloud as embodying 'magic'. Former *Rolling Stone* and *People Magazine* editor Alec Foege describes tinkerers as 'freethinking alchemists' who perform a 'peculiar strain of magic'. In *Tinker: The Art of Challenging the Status Quo* (2014), Glenn van Ekeren describes tinkering as 'the magic of momentum'.

The skill of an artisan is considered enchanted in many cultures. Many traditions imagine the artisan endowed with a *gift*, or else being accompanied by creative demons. Objects produced by artisans carry a magic aura, and the very idea of artisanship, involving *techne* and *craft*, has its etymological roots in sorcery, trickery and cunning, as I described in the previous chapter (recall, also, Ralph Waldo Emerson's belief that those who practise artisanal work 'by real cunning extorts from Nature its scepter'). Art, too, has a lineage of cunning, fakery and trickery – *artifice* is derived from the Latin *artificium* (workmanship), based on *ars* (art) and *facere* (craft, or make). The essence of a magical thing, wrote cultural studies scholar Steven Connor in 2000, 'is that it is more than a mere thing. We can do whatever we like to things, but magical things are things that we allow and expect to do things back to us.' To the anthropologist Alfred Gell, who spent much of his career studying magic in Trobriand Islander culture, this radical transformation, the metamorphic *wit* of making 'what is not out of what is', generates a perception of magic for the beholder and the maker (as magician) alike. To Gell, an object's magic is seated in its artisanal and technical *skills*: its very *making*.

So consider how you feel about the sourdough bread you bought from an artisanal baker, or the Yarra Valley wine you drank last night, or the thrillingly transformative book you're reading right now. Now consider the mechanisms by which you perceive these products as carrying some kind of aura that more mundane human products (say, a painted wall, a boiled egg) don't carry. To do this, it's helpful to adopt a ruthlessly rational distance that Gell described as *methodological philistinism*. I've simplified Gell's thesis into four easy steps:

Step 1: You have to devolve into a philistine; the kind of punter who'd embarrass art-literati types with: 'That's not art – my two-year-old could paint that.'

Step 2: Fully immersed in your philistine mode, you have to commit not to be 'taken-in' by the art. You must divorce yourself from ideas of inherent artistic value, power, quality, sacredness and authority – in short, its magic. This will free you up to understand art (and technique) anthropologically.

Step 3: To practise this philistinism, you need to abandon your sociological ideas of art (think: no Bourdieu) and your aesthetic ideas (think: no Panofsky), and understand art in terms of its *making*. You need to pay attention to the maker's *technique*.

Step 4: Just as your ancestors might have ridiculed the notion that art is a separate category from surgery, or from shoemaking and playwriting, you need to disavow accepted distinctions between practical and creative arts. You need to view art simply as a *made thing.*

When doing this himself, Gell found that art and technology are inseparable universally, because what we take to be art's 'magic' is simply technical skill. Once immersed in these steps, he could recognise art as the outcome of a technical process, the sort of technical process in which artists – not us philistines – are skilled. And so using this method Gell could argue that, anthropologically-speaking, engineers and surgeons are equally artists, and painters and songwriters are equally technicians – they are all *makers* and *performers*, just as they were in classical antiquity. To Gell, a major deficiency of the aesthetic approach is that art objects aren't the only aesthetically valued objects around – we value beautiful sunsets, too, and cute animals, and the exquisite patterns in a rock-form or tree-bark. But art objects are beautifully *made*, or *made* beautiful. (There's that ancient *ars* and *techne* again.) So it's the *skill* of the *made* bit, the technically-achieved human excellence, that gives an art object its magic. Magic is a product of 'the halo-effect of technical difficulty' that is made by a human but seen as superhuman.

To Gell, this asymmetry between the beholder's abilities and the perceived technical difficulty in making something explains why cultures not familiar with photography might perceive the photographer (and photograph) as magic. Conversely, some of us might consider photorealist painting in our own culture as more enchanting than photography, which is considered technically 'easier', or within our grasp. (Hence: *my two-year-old could do that.*) Gell believed that the moral significance we attribute to an artwork 'arises from the mismatch between the spectator's internal awareness of his own powers as an agent and the conception he forms of the powers possessed by the artist'. Since these technical difficulties transcend our own mastery of skills, we construe them as holding the magic aura we call *art.*

Chris Block in his largest shed, built for his boat-building adventures.

So the artist is an occult technician, an alchemist who through technical trickery transforms one thing into another. Even the magician him- or herself is spellbound by this process. When Gell studied Trobriand Islanders' magic canoes, those who made the canoes were equally enchanted by the canoes' powers as spectators were.

Powerful though it is, one problem with applying Gell's brutally reductive theory is that the tinkered object isn't always resolved or made well, and so, to some, part of tinkering's charm is akin to the charm we ascribe to naïve art: it's sometimes *forgiving* of shortcomings in skill, instead celebrating its humanist (rather than superhuman) spirit and adaptive *wits*. Gell's theory also doesn't explain conceptualism, nor how the most basic arrangements of banjo riffs, even when played with limited skill, can give us resonant goosebumps if they happen to twang with the right dissonant sweetness. Sometimes magic is something other than technical virtuosity. In my own enchantment of the iSlate, I was awestruck *before* I understood the technical feats its making involved. Making one wouldn't be too far beyond my grasp – or so I thought.

But when Block took me down the hill to the property's biggest shed, a vast space he built to accommodate his boat-building adventures, I was disabused of this notion when I noticed around a dozen discarded prototypes on a work-bench. Some of them had impurities in the slate; others split wood; some were crafted in the unfashionable profile of earlier-generation iPhones (each new model is edged with a distinctly different curvature). Clearly, many hours of painstaking trial and error had gone into perfecting them. And then, when he was explaining the mundane secrets of the iSlate's manufacture, Block revealed a heartfelt appreciation for things ergonomic. He spoke of how pleasing the iPhone form is in the hand; how Apple must have researched its curves, weight and shape; and that recapturing this shape in slate and cedar had increased his appreciation of it. He seemed attuned to the same humanising (or romanticising) sensibilities as *technium* philosopher Kevin Kelly, who described the evolutionary beauty embedded in everyday tools. Kelly wrote that in a good pair of scissors:

> the accumulated knowledge won over thousands of years of cutting is captured by the forged and polished shape of the scissor halves. Tiny twists in the metal hold that knowledge. While our lay minds can't decide why, we interpret that fossilized learning as beauty. It has less to do with smooth lines and more to do with smooth continuity of experience ... attractive scissors ... carry in their form the wisdom of their ancestors.

Before I met Block, Thomson had driven me to the South Australian town of Goolwa to interview John Yard, a tinsmith renowned for his rare-trade knowledge and skills. Because Yard doesn't use email, and he rarely answers the phone, he and I had exchanged correspondence by mail (he replied to my letters with his own, written in pencil). He told me his tools, once belonging to his tinsmith father, still transmitted the serendipity of his father's labour:

> It feels good when you hold them. There's a lot of history in these tools, a lot of work, and you know, a lot of love and good work went into them. They feel good, and they feel right, and it means a lot to me to use these, I do good work with these.

Likewise, when explaining the elusive idea of magic lying in *quality* objects and tools, Block told me that unlike poor-quality throwaway tools, he uses his for years and that 'Some of your tools, they become an extension of yourself. They're extensions of your fingers and hands. They become like limbs.' The idea of accumulated and projected experiences embedded within objects is echoed along a lineage of theorists including Ernst Kapp, Martin Heidegger, Marshall McLuhan, Bruno Latour and David Rothenberg. In his famous *Understanding Media: The Extensions of Man* (1966), McLuhan described technologies not simply as corporeal extensions, but as extensions of humanity. He described our tools as 'the technological simulation of consciousness, when the creative process of knowing will be collectively and corporately extended to the whole of human society, much as we have already extended our senses and our nerves'.

Three decades ago the philosopher Langdon Winner, building on earlier ideas by Lewis Mumford, explained how objects are political artefacts with our values embedded into their form and fabric. He described the unusually low-clearance bridges over parkways on Long Island, New York. Their structural peculiarity might seem romantic and charming, but Long Island historians made a shocking discovery: bureaucratic documentation showed that the bridges around the parklands were deliberately designed this way to achieve a social effect. Poor people and racial minorities, who normally used public transport, were discouraged from accessing the parklands because buses couldn't fit through these overpasses. The bridges were designed so the cultural elite could enjoy the 'public' parks free from the underclasses.

It seems unlikely that Winner would agree with Kelly's evolutionary idea of the *technium*, which doesn't exclude politics, but which describes technology as having 'near biological behaviour' and an organising culture 'like a very complex organism that follows its own urges'. This suggests an inevitable natural order. To Winner and others after him, there's a greater dose of political agency in the technium's genome than Kelly's theory suggests. Winner's example is disturbing, but his point is also explicit when we consider how tinkerers and hackers have given the world values such as 'open' (open source, open access and open software) from the bottom-up. These are values of hope. And capitalist, socialist and communist power structures alike have come to integrate them, however provisionally and imperfectly.

The tinkered project is the embodiment of hope, and tinkering's magic resides in a narrative of hope-defying-improbability (*what are the chances?*), and risk. The technical ingenuity of associative leaps (door mutates into iSlate, improbable cogs become museum display) could have *not* worked. That they did is magic and superhuman. Gell wrote that magic has been difficult to obtain because it 'requires a complex and chancy technical process'. When he studied the Trobriander carvers who craft canoe-boards in order to dazzle and weaken beholders, he asserted that the makers of the canoe-boards themselves are every bit as 'taken in' by their magic, because transforming the boards is so *technically* challenging. This technical challenge *itself* makes the canoe-board craft transcendent. The carvers undergo 'magical procedures which open up the channels of their minds so that the forms to be inscribed on the canoe-board will flow freely both in and out'. A beholder might be awed by the smooth, snake-like, sinuous lines of the board, and so perceive the board itself as enchanted. But from the carver's point of view, these lines represent technical difficulties that must be overcome with great mental capacity (for which he has been given specific magical rites).

For people who make, build and tinker with things, there's a point at which a material project becomes more than a sum of its labour and parts: it has its own palpable spirit and character that the maker couldn't foresee (or conjure). Novelists talk about involuntarily travelling to emotional and conceptual places their characters take them; musicians speak about the *esprit* that moves through them in an improvisation; builders and architects describe a point at which a building has asserted its own fluency, authority and charisma. Tinkerers, too, talk about being responsive to the beckonings of their materials. In all these forms of making, *ars* and *techne* are reiterated.

Even the habitually rational Gilda Civitico, a tinkerer profiled in the following chapter, described the kind of tacit instincts involved in her projects as an 'inexplicable thing' that can't be conveyed to someone 'without a feel for' the project. She described the enchantment of getting into 'the groove' of a project as if the project were its own discrete force, and the point at which a project transforms into 'its own thing', rather than the sum of her labour and material choices. Tinkerers are not simply skilled masters over their materials; their materials are equal collaborators, and the tinkerer keeps faith in a project that is bigger than her- or himself. Responding to questions about the iSlate's magic, Block told me:

> The shape itself has magic. The first one [iPhone model] was all curved, which was really nice, but bloody hard to do, sanding, and fiddlearsing around. Think about it, every aspect of something that is curved is hard, in terms of production. When you introduce that curve, you're wasting space, 'cause inside that curve is really just decorative surface treatment: that curve wasn't there for efficiency [in the original iPhone]. They probably just worked out that the curve *felt* better.

Fiddlearsing around. Just as Gell reckoned, the magic in Block's story is precisely what's so technically challenging and inefficient to achieve. The quantification of efficiency, productivity and innovation that dominates public conversation completely ignores the motivating force that drives our urges to produce. Even within the most practical, secular and unsentimental tinkerer's workshop, a sense of enchantment lured the tinkerers I met into their projects, and when their projects seemed improbable it held them in trance-like focus until the project was somehow, against the odds, resolved. When policy makers and bureaucrats make decisions about cultural resources like Men's Sheds, Hackerspaces, community gardens, manual education and innovation programs, they need to understand that their magic force lives in quests, stories, senses, skills and the plotting of self in the continuity of experience. Magic lives in hard-won skills, hard-to-find materials, moral fortitude, ancestral technium, and passionate spirit of tinkered projects; in stories, quests, serendipity – and *character*. For all its senseless and supernatural overtones, magic is an important way to understand the everyday transformative, spellbinding power that pulses through Australia's sheds, paddocks, kitchens, backyards and workshops.

4

Sanctuary

Some mornings, when Gilda Civitico arrived at the Victorian Infectious Diseases Reference Laboratory, where she worked as director of experimental design, she opened a cage of fluffy ducklings and eased them into a sink. 'I gave them a little pat and let them have a swim,' she recalls, 'with their mates'. Then she cupped one in her hand, anaesthetised it, and while its tiny heart still quivered, she sliced open its belly, snapped through its ribcage, disentangled the pulsing organs, and harvested its slippery liver, snipping off portal veins and connective vessels. 'You have to do it while it's alive,' she explained, 'or the blood clots'.

It's fiddly work, in which speed and precision are essential. You can't culture live cells if cell function starts shutting down. In this kind of experimental work, a confluence of material forces and knowledges interplay. You have to meld your theoretical knowledges (objective, measurable, explicatory) with your bodily knowledges (subjective, tacit, intuitive, skilful), and these are in continuous correspondence with the material forces around you (contingent, intractable) as well as the forces of time, climate and culture.

After harvesting a liver, Civitico quickly flushed out its red blood cells and then doused it with collagenase, an enzyme that breaks down the binding proteins and transforms the organ into a mound of substance the consistency of soft tofu. She pressed this slippery, still-warm mound into a sieve so fine that only single cells could find a passage through. The resulting slurry she spun slowly in a warm centrifuge, which separated the cells into three distinct bands: fat cells at the top; remnant red blood cells in the middle;

and, at the bottom, the denser cells she coveted – hepatocytes (the main liver tissue cells). A labyrinthine procedure followed, in which Civitico suspended those cells in prewarmed (37 degrees) culture media. With a process of gentle layering and centrifuging, she got the cell density just right. 'Think liquid red jelly being gently layered onto not quite set green jelly,' she told me. 'You don't want them to be mixed.'

This done, she syringed the substance into a pipette, washed the cells, counted them, and checked their viability by a staining process. 'And then you dilute them out with media to the right number of cells per ml of media and this is what you use to seed your cell culture plates.'

Left overnight, exiled in their Petri dishes and sustained with exacting temperatures and gas mixtures, the duckling liver cells bonded together and also to their new homes. And each day for the next few weeks Civitico tended to their needs, changing growth medium and nutrients. This microbial life-support regime, she explained, was but the first of many procedures before the finicky business of DNA profiling and dose-response analysis. Had she infected the ducklings with another strain of hepatitis B, or had she been testing a different viral inhibitor, she might have designed this experiment altogether differently.

Experimental science is an art form. It isn't merely a series of procedures. It requires fine-attunement to aesthetic detail and sensory attention. It's *ars* and *techne*. As with all forms of tinkering, experimental design involves educated guesswork and mindful interaction of procedural, tacit, bodily and propositional knowledges, interplayed with the knowledges embedded in technologies and materials. Just as a cellist might subtly adjust her technique to the proclivities and demands of a particular instrument, experimental science's raw materials and tools aren't simply passive recipients of

One of Gilda Civitico's dresses in progress.

human will, but active agents. It's the same deal from surgery to brick-making. Anthropologist Tim Ingold describes a confluence of human forces and materials in his description of clay brick-making, which 'results not from the *im*position of form onto matter but from the *contra*position of equal and opposed forces immanent in both the clay and the mould'.

In Civitico's lab, too, intimate attunement to these agents and their nuanced qualities – all their variabilities, tensions, fluxes, flows and resistances – was essential. Sometimes there was no telling what infinitesimally subtle variable of matter might violently sabotage a trial and undo months of evolving research. 'Sometimes,' she told me, 'a supplier would subtly change something. So for example, if you ordered foetal calf serum, you made sure you stocked up on supplies of the same batch. You want to minimise variables.' The right density of cell suspension, or the ideal nutrient profile in growth medium, might be knowable by keeping abreast of specialist literature. But other contingents – like the subtle colour- or shape-shifts that suggest pH change or cell fatigue, or the ways some cells seem to prefer certain plastic plates – were knowable through intense observation, tweaking, inkling and kinaesthetic knowledge. Some knowledge simply couldn't be codified in a procedures manual or even adequately explained to an assistant, because it relied on intuition. 'You get a *feel* for what the cells like,' Civitico explained. People would be surprised, she added, at how much tacit knowledge is involved in lab science. 'You can't explain it, but you just develop a feel for what works.'

In his celebrated essay on intellectual craftsmanship, the sociologist C Wright Mills argued that theory doesn't simply, through application, become praxis: instead theoretic knowledge is itself 'part of the practice of craft'. Craft – like tinkering – is a way of knowing by doing. To many craft theorists, the applied knowledge gained by making can't be disentangled from the theoretical knowledge behind that making. Sociologist Richard Sennett observes that 'all skills, even the most abstract, begin as bodily practices [… and then …] technical understanding develops through the powers of imagination'. There's a recursive relationship between making and thinking. In *Tacit Knowing* (1967), the chemist and philosopher Michael Polanyi asserted that humans carry a certain knowledge-awareness without being able to identify it in words. Craft has long been understood as ineffable: something learned with the whole body and its senses rather than simply the mind. As early as 1677, Craftsman Joseph Moxon wrote that craft 'cannot be taught by Words, but is only gained in Practice and Exercise'. In *The Art of the Maker* (1994), Peter Dormer asserts that the knowledge to make something work, or to understand how it works, 'is not the same as understanding the principle behind it'; tacit knowledge 'differs from propositional knowledge in that it cannot easily be articulated or described in words'. Polanyi distinguished between knowing *how* and knowing *that*; the former (the *how*) is being rapidly lost in a Western educational system that privileges more abstract vocational training geared towards information economies over manual, bodily or skills-based knowledge.

Although her former career as a research scientist gave her an understanding of the tacit knowledge of tinkering in the lab, Civitico's solitary tinkering in her home offered her additional freedoms: freedom to play and learn in a way that was unfettered by economic, political or bureaucratic concerns. In her home, tinkering became 'a playful pursuit', she told me. It's about 'applying knowledge you already have to a new problem or creative challenge. Tinkering is experimental so the results of your tinkering might be anything. I enjoy this element of uncertainty because I only ever have to please myself, and the process is always instructive.'

A few times when I visited Civitico, she and I sipped tea in the open-plan rear extension of the postwar brick home she shares with her partner, electrical engineer Andrew Peel (a tinkerer who was at the time researching a PhD in electrical engineering), and their two young daughters. Each time, her sewing machine and various sewing paraphernalia occupied half of the dining table. This was consistent across my research: tinkerers' living arrangements tended to be physically – as well as temporally – organised around their tinkering. In the light-filled space were shelves of games, shells, fossils and magazines: *New Scientist*, *IEEE Spectrum*, *New Economist*, *Make* and *Craft*, amid a collection of specialist books she calls her 'craft porn', including *The Art of Manipulating Fabric* and *Metal Clay: The Complete Guide*.

This library represented a more pleasurable stock of references than the procedures manuals she once followed in the lab; yet the deliberations and processes of a research scientist, Civitico told me, are not so far removed from those she now undergoes here, in her industrious domestic life. 'The things that made me a good scientist are what make me a good craftsperson,' she told me. 'I have a very high tolerance for repetitive stuff people find mind-numbingly tedious.' This repetition, she said, doesn't negate creativity, but enables it. Both lab work and her current occupation – as prolific maker of jewellery, clothing and preserves (the 'jam lady', as she was known at her daughters' school) – require strategic imaginings, a high frustration threshold and a willingness, dedication even, to learn from mistakes. She's happy to unpick a garment just as she might, in a lab trial, analyse a procedural error and start over again. 'Any kind of technical learning,' she says, 'requires you to research, imagine, plan, execute a technique, fail, troubleshoot, try again and keep tweaking'.

Civitico's story is one of psychological sanctuary – of tinkering-as-refuge. Her upbringing in an Italian-Australian family involved an array of cooking and crafting, but as an adult she rekindled crafting as a way to anchor her mind when postnatal depression hit hard. It was a way, she says, to channel mental and bodily energy into material problem-solving rather than the dark wrestle of abstract anxieties that besieged her. She could have returned to her job, but that meant dealing with other stresses and demands – bureaucratic, collegial, temporal, political – and it would mean neglecting what she considered the more important work of parenting. Nor did she seek the status or monetary rewards of work. She didn't especially *need* the objects she started making, though their material value (and social agency) became a happy side-effect. The refuge she needed can be understood as *engagement*, a way of transforming her mind-noise into a liminal rhythm that at times she'd experienced in her lab work.

Versions of engagement have been described in sociology, psychiatry, labour studies and neurology. Mills defined the craftsman as being:

> engaged in the work in and of itself; the satisfactions of working are their own reward; the details of daily labour are connected in the worker's mind to the end product; the worker can control his or her own actions at work; skill develops within the work process; work is connected to the freedom to experiment.

Building on Mills, Sennett defines engagement as the 'experimental rhythm of problem-solving and problem-finding' that craftspeople experience, while Crawford describes this rhythm as 'simultaneously technical and deliberative'. To Sennett, the 'carpenter, the lab technician and the conductor are all craftsmen, because they are dedicated to do good work for its own sake. Theirs is practical activity, but their labour is not simply a means to another end.' He essentialises the craftsman's rhythm. 'The craftsman,' he writes, 'represents the special human condition of being engaged'.

Engagement is recognised in contemporary psychology as 'flow', a term coined by US psychology academic Mihaly Csikszentmihalyi, who wrote, among other books, *Beyond Boredom and Anxiety: Experiencing Flow in Work and Play*, and *Creativity: Flow and the Psychology of Discovery and Invention*. In an attempt to codify the 'optimal experience' that gives life purpose, Csikszentmihalyi interviewed artists, chess players, and others whose work involved the rhythm of concurrent problem-finding and problem-solving. He found they could achieve a state of transcendental grace.

The way Civitico describes it, engagement or flow (she describes the state as 'a kind of groove, like a form of meditation') concurrently occupies the mind and quiets it. During this state it doesn't occur to the tinkerer to check her watch or to eat; one is 'completely caught up in what you're doing'. Many craftspeople attest to the allure of this rhythm. In an exhibition on rare trades presented by the National Museum of Australia, bookbinder Daphne Lera described:

> this feeling [that] happened almost within the first week of starting to learn bookbinding, and it hasn't really left me. It stayed with me all these years. It is to do with the fine physical task. It sounds repetitious … but it's also got this *rhythm* to it … this rhythm I'm talking about, all I can say is that I recognise it, and I know that it does, it does exist. I lose myself in it when I'm concentrating.

Lera says this happens when she is 'forever trying to perfect the technique'. John D'Alton, profiled in Chapter 8, described this liminal, meditative focus as 'a manifestation of God'. Polanyi, too, describes a kind of tacit human knowledge 'from which a harmonious view of thought and existence, rooted in the universe, seems to emerge'.

The way Civitico described it, entering this rhythm sometimes involves a certain discipline, pushing herself through a portal of frustration. She showed me a tube the size of a small child's finger. Inside were open-ended silver rings: impossibly small. Civitico told me her supplier, whom she sourced online, 'twists fine silver wire on to a mandrel, then she tumbles them to get the burrs off. When you choose them you have to be precise about the size of the internal hole compared to the gauge of the wire.' Once Civitico settled at her work desk with a pile of the right proportioned rings, she switched on her task-light, and took up two fine pliers. So began the painstaking process of opening, twisting and closing the tiny rings, and fashioning them into impossibly intricate patterns. 'There's a certain amount of getting the rhythm back – you're all thumbs,' she said. 'Sometimes you keep dropping them and swearing for half an hour before you get to that state, and then you could just keep going forever.'

This is as true of jewellery-making as it is of motorcycle repair. When fixing vintage motorcycles, Crawford, who described this material engagement as 'simultaneously technical and deliberative', reported:

> At this point I've exhausted my entire lexicon of 'motherfucker'-based idioms … and a certain calm takes over. I used to try and hypnotise myself into [this] Zen-like state of resignation at the outset. It doesn't work …

Seducing herself into a tinkering state helped restore Civitico's health in ways she never calculated. She started tinkering from home by chance. Late one morning, as she placed a necklace – a birthday gift from her brother – on her dresser, she paused. 'I looked at it and thought, "I can make this".' She hadn't really considered how jewellery 'works', but as she took the time to consider it closely, with the trained eye of a microbiologist, the necklace revealed its workings to her. It was 'made of a piece of flexible coated silver coloured jewellery wire … onto which oval shaped, flattened beads had been fixed at regular intervals and held in place by small crimps. It was closed with a spring clip and tag … when I had a really close look I thought: "How hard can that have been?" I wondered about the crimps and how they stayed in place.' So she jumped online, and a universe of adventure and possibility revealed itself. It was a *coup de foudre*. Once she had the tools and materials, she made the necklace successfully and, recognising her handiwork as the neophyte impulse it was, she unpicked it, and set out to make something more challenging.

Civitico's craft room, which doubles as the family's music room, houses a floor-to-ceiling wall closet that itself houses a hierarchic organisation of boxes within boxes, like a monumental Matryoshka doll. Some of these contain buttons, folded vintage fabrics, patterns, and tiny tubes of infinitesimally small jewellery components. There are varying grades of wire (with names like 'half-hard' and 'tigertail') and silver coils. There are regular- and irregular-shaped beads (these have names like 'bugle', 'hex', 'Charlotte', 'seed', 'faceted' and 'Japanese delica'). And handmade lampwork beads she ordered from the US, inside of which jellyfish-like forms are suspended. 'People slave over a hot torch flame to produce these,' she told me, 'from coloured glass rods'. (To be sure, if you Google 'lampworking', you'll discover a vast lexicon of technique and tradition.) In each tiny globe was a universe of otherworldly forms. We squinted for a few moments, holding each of them to the light and marveling at their innards:

nacreous, opalescent and ethereal sea-floor forms. 'Part of the joy of craft,' Civitico said, 'is the joy of discovery. It awakens in you the possibility of things, and whole other worlds.'

Tinkering's intensely insular practice can be experienced as expansive. Mark Thomson (profiled in Chapter 9) described tinkering as 'an intense focus on a small world', but, like meditative practice, this micro-focus can transport you to unbounded spaces. It can also unpick and completely dismantle a Cartesian triptych of formal categorisations that routinely accompany our understandings of materiality and happiness: *having, doing and being*. In his seminal *Being and Nothingness* (1943), Jean-Paul Sartre built upon Descartes, Husserl, and Hegel to introduce these three 'cardinal categories of human reality'. Later, in David Hume's *Treatise* are questions of object *as opposed to* experience. The way tinkerers describe it, the hypnotic, insular state of material engagement integrates all these elements (object and experience, having, doing and being), in ways that can be transcendent.

Civitico's collection – her stitching patterns, her beads, her weaving manuals – bear strong reference to microbial forms, and also cosmic forms. A bead structure resembles the hallmark corkscrew coils of proteins; a necklace replicates exactly the twisted bands of DNA; there are versions of mandalas that can be found everywhere in nature; there are asteroids and supernovas. We moved on to onyx flowers and military buttons. Civitico unfolded some waxed paper to reveal black Victorian glass buttons she had bought online and excitedly awaited. The very first time she edged them out of their packaging, 'I gasped,' she recalls. She told me she might choose to brick-stitch these treasures, or use a wide-angled weave, or employ square-stitch or peyote – there were any number of traditional ways of working jewellery in her collection of craft porn. 'You choose your torture,' she said.

Civitico's home is a Tardis. Each time I visited she disappeared down the corridor and returned with a trove of handcrafts of a scale seemingly disproportionate to the capacity of the house. There were French linen dresses and coats crafted from patterns she made by buying artisanal clothes from Etsy, unpicking them, working out their engineering, and modifying them to her own designs. (In tinkering language, she'd reverse-engineered them.) There were intricate charm bracelets and beaded necklaces. She showed me a chainmail bracelet and I involuntarily caught my breath: thousands of tiny silver coils linked together to form a wide, flat band whose elegance belied the finicky, curse-ridden labour it embodied. Its weave had the uniformity of machine-knitting. Many such chainmail bracelets might appear in any number of chain-store jewellers in countless impersonal shopping centres, and in this context such a chainmail bracelet would seem insignificantly generic to me. Had I seen this one in the context of a shop window, I would never have considered its workmanship, but would have passed it by without thought. I told Civitico this. 'We are over-exposed to beautiful things,' she answered.

Culturally saturated in a cornucopia of mass-produced things, we've become desensitised. In *Vibrant Matter: A Political Ecology of Things* (2010), political theorist Jane Bennett describes what this does to us. She believes that the 'sheer volume of commodities, and the hyperconsumptive necessity of junking them to make room for new ones, conceals the vitality of matter'. Matter's vitality is concealed not just by its ubiquity, but by our wilful obliviousness to the global economy's dark pockets. Much of our stuff is dumped in the very regions where it was made by underpaid, marginalised people and inhumane, toxic working conditions. Shipping containers arrive at our shores to deliver our consumer products, returning not empty, but full of our material waste for more marginalised people to sort through. According to Adam Minter's *Junkyard Planet* (2014), the global trade in junk is the world's largest employer after agriculture, with an estimated 65 billion tonnes of discarded stuff channelled to various scrapheaps. We enjoy a comfortable distance between ourselves and the high-volume, low-margin scrapyard economy, its junk-lords and its rag-pickers. In some aspects, waste-harvesting is the global economy's most polluting and exploitative industry; in others it's our most sustainable and hopeful. Capitalist, socialist and communist societies alike – wherever commodity-products are manufactured and consumed – are complicit in this economy.

The other side of our concealment is our own dissonant condition that psychologists call *hedonic adaptation* or the *hedonic treadmill*. We love new things, but these theories posit that our material acquisitions give us temporary pleasure that is eventually subsumed into everyday life and taken for granted. We withdraw our love. Once they cease to provoke happiness, we set out to accumulate more material objects to love. Within psychology literature, the hedonistic happiness enacted by material consumption is characterised as a shallow and temporary pleasure, keeping people apace of superficial happiness without deep and sustainable enrichment. By contrast, *eudemonic happiness* is defined as a deep contentment.

There are parallel theories of design philosophy and marketing. Design philosopher Jonathan Chapman asserts that utopian fictions are embodied in consumer products, but once these products' fictions are demystified or outmoded by other utopian ideals, they are discarded and new products are sought. In their vastly-cited 2007 and 2010 studies of young adults' relationships between materialism and self-esteem, marketing academics Lan Chaplin and Deborah John found that people try to build their status and identities by accumulating certain kinds of material objects, but that those of us with high self-esteem are less likely to be materialistic in this way. A large body of psychology research has subsequently linked commodity consumption with low self-esteem and character traits of envy.

But tinkering unravels the understanding of materialism as a form of commodity consumption. Instead, tinkerers have a deep and active personal and social engagement with material objects. Many psychology, marketing and industrial design scholars view production and consumption of goods more as an expression of capitalism than an expression of humanity, but other scholarly disciplines look outside of the commodity-object lens. Miller and Csikszentmihalyi assess materiality instead as the depth of people's everyday active engagement with objects – regardless of whether these objects are produced or consumed.

Tinkering occupies an ambiguous place between production and consumption. Anthropologists including David Graeber, Jonathan Friedman and Marshall Sahlins have questioned the positioning of consumption in opposition to production in late modernity, arguing that this dialectic serves merely as a function of an economic schema. And recognising that household production contributes to informal economies, economists have started to reframe household 'consumer products' as 'household capital'. Purchasing a guitar might be categorised as consumption; playing it might be categorised as production, while much digital activity can be characterised as both. Both Miller and Csikszentmihalyi found that the more profound a person's material engagement, the more likely they are to be fulfilled in the eudemonic sense (as opposed to the hedonic sense). If we consider Chapman's idea of short-lived utopian narratives embodied in consumer products (discussed further in Chapter 8), we can see how tinkerers have deep and sustaining narratives in the objects they produce and relate to. They have a deeper fulfilment from materiality than commodity-object understandings suggest. To Chapman, for a product to have long-term emotional endurance, it must 'possess richer, lengthier and more complex fictions if the consumption process is to be satisfying and longer-lasting'. Meaning-making engagements with products increase their 'emotional durability' and thus their sustainability – waste being a symptom of 'expired empathy'.

In Civitico's home, I was confronted with the bracelet's rude materiality, its status as handcraft, its relationship with the maker. Despite the countless similar bracelets set in commercial contexts (jewellery chain-stores, advertising, catalogues), all of which routinely escaped my attention, this uncommodified one in the here-and-now seemed fascinating and exotically defamiliarised. Now that I wilfully shed ignorance and took notice, its presence was something other than a sum of its labour, parts and time. It was simply its own beautiful thing. Civitico told me she can recognise the bracelet's independent thingness, but she can't relate to her handcrafts as entities separate from her labour until after she has secreted them away, sometimes for a couple of months, herself defamiliarising and mystifying them, before retrieving them afresh and appreciating their craftsmanship and beauty. Perhaps this is her private inversion of hedonic adaptation.

Civitico's jewellery isn't commodified: instead, she gives it away. She gets the odd paid crafting commission, but she tells me: 'I can tinker for my own pleasure, but if I had to do it as a paid gig, it would get very stressful.' And yet she rejected my suggestion that her occupation is a hobby, practised as it is in domestic time without payment. 'No. Collecting is a hobby. Making stuff isn't.' Is it leisure?

'No. Leisure is passive. This is work, but with a different purpose and focus.' Many tinkerers describe tinkering as work that isn't toil. Civitico's response bristled at the words *leisure* and *hobbies*, as if these categories were insufferably bourgeois and pointless pastimes. Her tinkering was an *occupation*, she said, a word usually associated with formal work. It had *purpose*.

The problem of regarding hobbies and leisure as passive pastimes in dialectic opposition to work was famously introduced by French sociologist Henri Lefebve in his 1958 essay 'Work and Leisure in Everyday Life', which built on Marxist ideas of alienation. Lefebve argued that the centrality of formal work to people's everyday lives means that time at home is conceived not as 'a foundation of personal development', but instead as a compensatory *break* from work. To many Australians, home time is temporally and spatially organised by work, and hobbies are regarded alongside entertainment as passive and transient distractions. (Work tends to be productive; leisure tends to be consumptive.) Lefebve believed the result of this fissure is a fragmented selfhood, and that leisure routines away from the workplace can be as alienating as routines at work, and not a source of purpose and fulfilment. This insight is especially instructive because tinkerers reject the centrality of external (or formal) work to life – including work's spatial and temporal organisation of other categories of life (such as sleep and leisure). To tinkerers, leisure isn't time spent recuperating from (and for) the important business of work. Instead, it is time for human vitality and self-knowledge: in both the autodidactic and psychological senses.

The common distinction between pleasure and necessity is also defused in the tinkering mindset. Jess McCaughey, a prolific maker of artisanal soft-toys, described her domestic craft as not simply restorative or therapeutic: 'I get annoyed when people say craft is something no longer done out of necessity. I *have* to make things. It's not a luxury for me.' Instead, McCaughey regarded her home-based craft as sanctuary *and* vocation – essential to her sense of purpose. Similarly, tinkerer Cliff Overton told me he identified with his steampunk practice at home in Greensborough more than with his waged time outside the home as an industrial designer and firefighter. This understanding challenges common discourses around a 'work–life balance': the idea that work is somehow oppositional to life and upsets equilibrium, which needs to be compensated for during leisure time. Clearly, tinkerers see their tinkering as life, work and leisure.

So we come to the question of economic privilege. Who among us has the luxury to disengage from the formal labour market? It would be easy to suppose, from this book's bias towards white educated home-owning males, that tinkering is a privileged expression of the moneyed classes, a luxury pursuit of those who can afford to have *time on their hands*: leisure time, hobby time – *free* time. The Australia Institute's 2003 survey into reasons why some Australians opt for a lower-income, less consumptive, more sustainable and balanced lifestyle shows that 24 per cent had no work income, and most had incomes well below or just at the national average. Blue-collar workers were more likely to stop external paid work altogether, while white-collar workers were more likely to shift careers – but even so, the authors report, 'The attraction of downshifting is not confined to any demographic group.' Chapter 5 describes many studies that show how productive home enterprise is not necessarily linked to economic privilege – it's more likely to be linked to rich social connections and concurrent urges for individual autonomy (as it turns out, these aren't in opposition).

And there are many ways the tinkerer's home industry enables alternative systems of exchange that connect tinkerers socially but disconnect them from the formal economy – freeing them from the necessity of a formal wage. Civitico's family, for example, pays little GST, because her tinkering work (including clothing production and prolific cooking) limits their need to purchase processed goods. That the family was in a position for her to decide against paid employment also wiped out a wad of income tax. The materials she uses are not mass-produced DIY products of the kind that might be purchased at retail outlets like Bunnings, but very often old or remnant materials, specialist components produced by other cottage industrialists, or primary produce from friends – and thus they are exchanged rather than purchased. The products she makes are integral to an informal exchange ethic she calls her 'black market of jam'. Gifts are made, not bought; there's jam for the woman at the fabric remnants store; there's reciprocity between friends and colleagues; there's exchange for produce or sometimes goods from other home industries. Parents from school know her from her products and request them. This exchange is not tallied by a formal calculus. The relationships enabled by her craft attest to the seemingly outdated Marxist and anarchist ideas that a sense of selfhood and social relations develop through making physical things.

At the heart of the social contract is our payment of taxes in exchange for public services; and the quoins of our capitalist structure are the surplus value of our labour and consumption in our leisure. But when market-economies' contracts and conditions are by increments eroded in ways that undermine our humanity, we reinvent our own, humane, stealth contracts. For many tinkerers, versions of Civitico's 'black market of jam' – a certain sanctuary from *quid pro quo* ideals of exchange – can maintain many of the important human benefits that modernity is thought to have hauled away:

community, social relationships, and our senses of meaning, agency and autonomy.

The story of Benalla-based tinkerer Michael Drinkwater, elaborated upon in other chapters, offers an excellent example of how tinkering creates grey economies. In converting a dishwasher into a biofuel processor that served all his energy needs, Drinkwater bypassed the formal economy and in doing so fostered social and economic exchange that can't be formally tallied. First, he accessed open-source information to find out how to convert used restaurant oil into biodiesel. He found the ingredients (used vegetable oil, ethanol, caustic soda) were cheap or free: 'I had a deal with the garbage contractor at Mt Hotham and would take all the used oil, usually five or more tonne, from him at the end of winter [in exchange] for two slabs of beer.' In making his own biofuel and glycerine byproduct, he avoided commercial fuel consumption and production, and he created his own informal economy, giving the glycerine byproduct to his mechanic, who found it was: 'a great hand cleaner and degreaser. My mechanic couldn't get enough, and totally biodegradable.' In turn, the mechanic 'gave me motor oil for the glycerine,' even though Drinkwater emphasised this wasn't done according to a quid-pro-quo calculus: he and the mechanic 'did many things to help each other without really keeping a score sheet'. Instead, such exchanges, like those of Civitico's, became built on another register of goodwill. This register of goodwill, increasingly rare in the formal market-economy, was a consistent feature among the tinkerers I studied.

While Civitico's work isn't always consciously motivated by an anti-consumption or alternative-economy ethic, the effects of her industry coincide happily with her values. 'I cook according to my conscience,' she told me. For an hour and a half I stood with my notebook in her kitchen at the home's centre, watching Civitico, svelte in knee-high boots, fitted jeans and a large apron, make jars of quince jelly that would then be given away to family, friends and colleagues. Holding a jar of one of the previous day's batches up to the light, she told me quinces 'are the most underrated fruit. You could *never* find this in the supermarket.' Making it to this standard was a convoluted process, for which hundreds of jars were stacked in the hallway. In boxes were some of the 500 or so fruit from the half-century-old quince tree growing on her in-laws' grain farm near Winchelsea. The evening before, she stewed 14 for three hours, until the water had turned magenta, and she left them overnight to drain slowly through fine muslin into a bowl. 'If you try to get extra liquid by squeezing them, you'll pay later with sludge. You can't speed this process up.' 'Slow' is a standard theme among tinkerers; the conviction that to do a job adequately takes committed time – time to fail, time to engage, time to learn and get a feel for the project.

After I arrived, Civitico set a dozen jars in the oven to sterilise. She placed a handful of clean teaspoons in the freezer, measured the quince syrup and worked out the required sugar with a calculator (450g sugar for every 600g of liquid). She squeezed two lemons and poured the juice and the syrup into a stainless steel pot, adding the sugar and bringing it to the boil. It was difficult to imagine how the process that followed could be mechanised or rationalised on an assembly-line. It simply demanded too much sensual attentiveness, it was too humanly involved. 'You have to pay attention,' Civitico explained, 'to how many bubbles form and how slowly they come up'. She stirred with a practised hand, every few minutes taking up a flat straining spoon, removing the frothy scum that formed on the surface edge, washing the strainer and returning to stir. This process – stirring, straining, washing, stirring – continued for around

40 minutes. 'I like this bit,' she smiled, and her contentment was palpable. 'As it reduces down,' she showed me, 'the colour intensifies and the syrup becomes a lot clearer'. It was evident: the syrup had dramatically transformed from a musty magenta to a glassy ruby; its surface was mesmerising as it played with the light. Civitico told me she often gauges whether a jam is ready 'by the way the spoon moves across the bottom of the pan – the gap you make and the way the spoon moves into the trough'. But quince jelly has its own proclivities: 'See how it's reducing – there's the remnant froth, there's the tide mark around the pot.' She kept a rhythm of stirring, straining, washing, stirring, and we were comfortable just contemplating, speaking only when a change in the substance provoked a remark: 'The bubbles are getting slower, and now it smells different. It's changing density. You just watch it and wait. It will do its own thing when it's ready.'

The kitchen was now fragrant with tangy humidity, and we waited and watched, relying on our senses to alert us to when the syrup decided to undergo its next transformation. Civitico's collaborative approach with her materials – *It will do its own thing when it's ready* – iterates Crawford's observation that one of the values of material engagement as a way of understanding the world is that it involves wrestling with the limits of your own mastery over things, giving us the tools to understand the world materially, and hence critically. To Crawford, making and repairing things teaches us to 'submit to things that have their own intractable ways'. Materials don't simply succumb to our will: they equally demand we succumb to theirs, and their recalcitrant qualities give them agency and control in the making process that's instructive to our humanity. Ingold describes his students attempting to weave traditional baskets in collaboration with the forces of wind and those of willow sticks. Each stick 'did not want to be bent into shape … it put up a fight, springing back and striking the weaver in the face. One had to be careful and coaxing. Then we realised that it was actually this resistance, the friction set up by branches bent forcibly against each other that held the whole construction together.' So these forces, as much as cultural ones, had a role in the vernacular of basket-weaving.

Crawford believes there are moral lessons of materials' recalcitrant ways. (The geophysical backlash that is climate change is a dramatic example of this.) These lessons allow active, mindful engagement rather than passive, distracted consumption. To Crawford, if we approach the world with a lack of involvement with material production, gaining little material literacy, then we can feel less responsible for it, and thus less accountable, instead living our lives 'in these channels directed from afar by these vast impersonal forces that we don't understand … that's why tinkering is important'. He, like a lineage of craft-reformers before him, believes in 'a moral education tacit in material culture'. The learned helplessness that comes in the absence of material engagement 'leaves us bereft of something at the core of being human. And that is individual agency.' Himself a motorcycle mechanic as well as a philosopher, Crawford has his literary comrade in Robert Pirsig, author of *Zen and the Art of Motorcycle Maintenance* (1974), who writes:

> The place to improve the world is first in one's own heart and head and hands, and then work outwards from there. Other people can talk about how to expand the destiny of mankind. I just want to talk about how to fix a motorcycle. I think that what I have to say has more lasting value.

In Civitico's former work as a research scientist, agency over the material gave her this sense of moral responsibility. She didn't delegate manual work to lab technicians because: 'if there's a cock-up it's mine. I don't want to live with someone else's mistake. Besides, you're more likely to understand what went wrong if you made the mistake.'

Ideas linking material production and morality are deeply embedded in European and Asian cultures. According to Sennett, early Christianity placed great significance on Christ being the son of a carpenter. From the Middle Ages, craftsmen-saints appeared, and throughout Pagan and Christian doctrine 'is the belief that the work of one's hands can reveal much about the soul'. (Once again, recall Ralph Waldo Emerson's assertion that DIY work is 'God's education'.) Material production was a way to prevent idleness, a sinful temptation in early Christian doctrine that persisted with secularis-

ation and industrialisation. But as materialism became equated with consumption, Protestant, anti-consumption and anti-obsolescent ethics saw it equated with moral failure. The European adoption of various Asian doctrines including Buddhism, Hinduism and Jainism have further reinforced anti-materialist ethics. These doctrines, according to Miller 'took a much more profound interest in the centrality of desire and materialism to the condition of humanity and its relationship to the world than did Judaism, Christian or classical teaching ... in India the avoidance of materialism ... became essential to the quest for spiritual enlightenment'. But if we look outside the consumption model, we can see how a communion of human and material forces can offer a two-way secular transubstantiation. The made thing is transformed by the human; the human is transformed by the made thing.

After some time in Civitico's humid kitchen, the bubbling became finer and the whole pot started to roar. 'Here it goes,' Civitico said.

Her pace quickened as she began dipping the teaspoons from the freezer into the syrup and dragging her finger along the film of jelly that formed on their backs. 'When the tracks become more distinct you know it's about ready.' She was intensely focused, concurrently judging the films of jelly on the back of the teaspoons against the tensile stringiness with which the syrup dripped off the wooden spoon. 'You can't just go by one thing: there's more than one marker.' There are many factors indicating a jelly's readiness. If the jelly is removed from the heat too soon it will not set well; if removed too late it will spoil. Here is the trickery involved in transformation.

'Here comes the bit that shits me,' Civitico warned. The bubbles rose dramatically higher up the sides of the pot and the surface foamed furiously. 'Right at the crucial moment,' she said, 'it starts to raise all this stuff'. The foam, she speculated, is 'probably a protein precipitate or a complex polysaccharide'. Once again, her theoretical knowledges (objective, measurable, explicatory) and bodily knowledges (subjective, tacit, intuitive, skilful) corresponded with material forces (contingent, intractable) as well as the forces of time, climate and culture.

Suddenly: 'It's there!' After a quick scum removal, Civitico employed a practised technique of carefully pouring the syrup down the inside surface of each jar. 'You do it this way because you don't want to introduce bubbles.' And there they were, nine-and-a-half jars of pristine, deep ruby-coloured jellies, more than a day after the process started. (Longer, if you take into account planning jars, harvesting fruit, writing labels.) It's a slow but triumphant success, and after our intense focus on this small world, we marvelled at the colour and jewel-like purity. Giving quince products away, Civitico told me, 'is an act of love, because it's a pain in the arse to make'. Love and social exchange, then, was bound with technical feats. She then gave me a jar and some paste as a gift. After momentary hesitation (should I accept a gift from a research participant?) I felt both elated and indebted, inducted as I was into the black market of jam.

5

Home

John Tucker told me that you can master any material project if you just do the sums and then, you know, clamp that bit there and shave off this bit here, and make sure those two metals have something insulating between them so the electrolytes in this one doesn't make the other one corrode. In his flannelette shirt, overalls and a cap, Tucker spoke in the kind of softly melodic cadence you'd expect from a Sunday classical radio broadcaster. He considered few DIY projects beyond his skill-base or his capacity to learn, and, like other tinkerers, he freely dropped theoretical knowledge into practical explanations, frequently overestimating my own material literacy.

I met Tucker, a tertiary-trained musician, through his wife Kerrie, with whom I had contact through my advocacy work for an environmental NGO. (Kerrie is a former ACT Greens parliamentarian.) The couple had recently relocated from the ACT to Mount Toolebewong in Eastern Victoria, to be closer to their three adult children and young grandchildren. Tucker had built two of the family's previous homes largely from salvaged materials and help-at-hand, but this time the couple bought someone else's unfinished house-build at a good price.

So Tucker was a homemaker – an inappropriately gendered term. He'd built all his homes, not as a qualified professional but as an owner-builder. Although house-building was his main focus, Tucker also tinkered in photography, instruments and recordings. Kerrie, too, was a maker, mainly of mosaicked outdoor sculptures that were installed around the property. She'd assumed responsibility for the home's comforting details – vintage fittings from junk shops; handcrafts by Jessie; sculptures and veggie beds in the garden; contemporary paintings passed from Kate – alongside her paid research work that fitted between all these. Like many tinkerers I interviewed, the couple's material habits had transferred to their children's vocations: Kate is a successful painter; Jessie is a well-known maker, illustrator and textile designer; and Christian is a maker and industrial designer.

If we revisit early ideas of making, such a family is easy to recognise as tinkerers. Anthropologists have documented the familial and social ways craft is shared and passed on in many societies, with precedents reaching back to ancient Greece, when the idea of material skills being transmitted from one generation to the next was taken for granted. According to cultural studies scholar Jonathan Sterne, Aristotle viewed *techne* as both the process of making things, and the knowledge that allows for that production. (Sterne sees parallels between this understanding and Marx's famous adage that people make history, but in conditions not of their own making.) According to historian Larry Shiner, ancient definitions tended not to distinguish between our engagements with material forms, creative habits or even abstract design. Welding a pipe; playing an instrument; fixing an engine; cooking jam; moulding a sculpture; making biofuel; telling a story – ancient ideas of the maker's *procedure* didn't distinguish between raw materials of a play, a shoe or a medical technique. Each form's essential concerns were considered the same. Playwrights, shoemakers and medical practitioners alike were considered makers in classical antiquity. Fictional characters and medicines alike were *raw materials* for procedures in the same ways as leather or wood were. According to Sennett, Anton Chekhov understood this when he used the word *mastersvo* to describe both his craft as a doctor and his craft as a writer – and accordingly, ideas of art and craft *practice* emerged from the same roots as

John Tucker completes the fire-resistant balcony at the southern side of the house-build.

practical and *practitioner*. Using these ideas, it becomes impossible to differentiate between the useful and the creative domestic routines of everyday life.

The first time I visited, the Tuckers' new home-in-the-making comprised a concrete slab supporting the foundations of a passive energy-efficient bushfire-resistant house; an established orchard; a large fenced vegetable garden; a shed; a chook-pen (recently built by Tucker); and natural woodlands leading down to a majestic valley view. The property had mains power connection, solar power and tank water. It had landline phone and internet connection. To save money, the Tuckers would live roughly (and without legal approval) inside the house-build while it was under construction. Although Kerrie had some income from her part-time research work with The Australia Institute, Tucker didn't yet have paid work locally, and he would work on the house while he put his feelers out for paid opportunities.

If you're a home-based tinkerer, you're choosing a precarious life whose source of security lies in your own accumulated skills, social connections and adaptive wits. In *The Practice of Everyday Life* (1988), sociologist Michel de Certeau theorised that the *bricoleur* (tinkerer) uses tactics, rather than strategy, to 'make do' in everyday life. Borrowing from military theory, de Certeau described strategy as the technique of those in positions of power, but strategy's weakness is that it presumes control, management and predictable outcomes. Tactics, on the other hand, involve the here-and-now lived reality of those at ground-level – the make-do, adaptive, on-the-fly techniques of the non-powerful. While tactics are agile in the face of unpredictability, strategy is undermined by unpredictable events or elements. In de Certeau's understanding, the everyday deployment of tactics is *bricolage* – tinkering, or the art of making-do. Unlike strategy, *bricolage* responds to change in imperfect situations and is constantly reoriented for the situation at hand.

Under current conditions, these tactics involve freedoms and choices that aren't available to every Australian – especially not to those trying to scrape through on subsistence income. Many tinkerers I spoke with viewed these choices and freedoms as a basic civil right, if not a human right (see Chapter 10). For better or for worse, if you're a tinkerer, you're choosing to live opportunistically and by your wits, taking up whatever contract or commission happens to come your way or works amid your various projects. You get by within informal economies, and from time to time you accept paid jobs, temporarily relinquishing your freedom and autonomy by renting out your skills. As challenging as the tinkering life can be, it's a way to set your own terms and autonomy; no employer can dictate your time routines and buy your life outright.

A year into my interviews with him, Tucker had supported his house-building work by establishing some local music-teaching work, augmented by Kerrie's research income. He'd also secured some income from making curries – a passion – for guests at a neighbouring bed and breakfast. Although he was saving money by building his own house, tinkerers will tell you that the value of tinkering isn't purely economic. On the other hand, they'll also tell you thrift is a central concern of tinkering. Many of the 32 people I initially interviewed had never opted for high-income, high-consumptive lifestyles (and so weren't 'downshifters'), and Thomson's prolific research in the popular-market *Blokes and Sheds* and *Makers, Breakers & Fixers* depicts a spectrum of tinkering that knows no class or occupational boundaries – it's practised across the socioeconomic and ideological spectrum. 'All sorts of people, rich and poor, city and country': those from all walks of life feature in these books – the only exception being the very homeless or underclass that the 'tinker' origin denotes.

Scholars in the fields of economics, sociology and anthropology have accrued powerful evidence that DIY isn't generally motivated by thrift alone. In *Serious Leisure* (2008), sociologist Robert Stebbins finds that people are attracted to project-based leisure because 'they find irresistible its core activity.' A 2004 study by economist Colin C Williams found that 'do it yourself (DIY) consumers [people buying DIY products] have emphasised human agency rather than economic constraints when explaining their motives'. A richer story still emerges in Williams's later book *The Hidden Enterprise Culture*, which examines undeclared work practices in informal cash economies. In this study, neither affluence nor poverty are found to be reliable predictors of informal work. Williams finds no unicausal or generalisable factor across cultures, although educated people tend to be more likely to engage in *autonomous* informal work. Similarly, in their 2007 study of DIY renovators' relationship to materiality, British sociologist Elizabeth Shove and her colleagues found that many DIY renovators 'had the means to but were unwilling or unable to identify and pay someone else to produce the distinctive and innovative solutions to which they aspired and which they could achieve themselves'. Here, the story is not just one of individual agency and adventure, but a belief that amateurs will do a better job than professionals.

More recent work by Williams has documented a shift in post-industrial work economies. His research shows that 'work from the unpaid to the paid sphere has not only stalled over the past 40 years but in some nations it has even gone into reverse'. In other words, formally-paid work is in decline, and informal unpaid work is increasing. In the last decade, economists have recognised that more than half of all productive activity in industrialised countries happens in the home (if we tally cooking, child-care, repair and maintenance, photography, social media output, renovation, handcrafts and so on). Yet this productivity is not counted in GDP. Even so, the home is starting to be recognised as an important site of social and human capital. The work of feminist economists especially acknowledges that home production is an overlooked contributor to macroeconomic activity and social cohesion. Australian economists including the late Hugh Stretton have called for some household 'consumer products' to be reframed as 'household capital'. As economic geographers are shifting to more gender-inclusive and pluralist foci, and casting their gazes on informal economic exchange and production, they're observing concomitant shifts in how developed nations understand their modes of work.

In *Stuff* (2010) and *Materiality* (2005), British anthropologist Daniel Miller found that a commitment to material engagement doesn't have a straightforward relationship with income, class or consumption. In his study of Londoners inhabiting reviled council flats following the Thatcher years, the residents who renovated, updated, modified and decorated their homes didn't tend to be those with more disposable income or status, but instead those who had strong and fulfilling social relationships. People who transformed their material environments did so with the practical and idealistic influences of their social resources. Those with few and shallow social relationships were least likely to materially transform their council flats. This finding, wrote Miller, 'was the very opposite of that common assumption made in accusation of materialism, a word that implies that people who become focussed on their relationships with things also do so at the expense of their relationships with other people'. In Miller's case-studies, the home itself counts as a form of material production, and people's relationships with their homes show that 'usually the closer our relationships are with objects, the closer our relationships are with people'. His work affirms Marx's famous proposition that material production is simultaneously the production of people and social relations.

Crawford believes that a renewed interest in self-reliance and material competence with everyday materials and technologies – car engines, appliance innards, hand-tools – 'seems to have arisen before the specter of hard times' in post-industrialist economies. 'Frugality may be only a thin economic rationalization for a movement that really answers a deeper need. We want to feel that our world is intelligible, so we can be responsible for it.'

On my second visit to the Tuckers' mountain home-in-the-making, Kerrie told me that as much as she loved the property, it was yet to feel like home. We speculated that it might feel like home once living routines (harvesting, cooking, making, socialising) overtook the construction routines. In his ethnography *At Home in the World*, the Australian anthropologist Michael Jackson writes that his experience of the Warlpiri people 'business camp' (in the Northern Territory) 'reinforced my conviction that home is grounded less in a place and more in the activity that occurs in the place'. Jackson's informants were nomadic, so his research question was: do these people have a sense of home? He found the Warlpiri people experience their feeling of home through the activities, habits and stories of generations. While the tinkerers I studied don't have this kind of intergenerational relationship with their physical homes, they do tell a consistent story in which tinkering *itself* figures as a personal, familial and cultural continuum that's inseparable from the idea of home. When the home becomes a site of tinkering, tinkering itself becomes a form of home-making.

So if 'home' is as much symbolic and emotive as physical, then considering the home as a 'site' of tinkering involves thinking beyond a location and into temporal, imaginative, emotional and relational spaces. In modern Western imaginaries, home is conceived as a place of privacy, return, origin and retreat, and a repository for meaningful belongings and intimate relationships with people and things. In *The Comfort of Things* (2008), Miller affirms that the homes with the most material belongings aren't necessarily those with higher consumption habits or those with status-anxiety. The most disturbing of Miller's case-studies of Londoners' homes is that of a man named George, who has very little social contact, is biding his time until his death, and who has a brutal austerity in his home: no ornaments, no souvenirs, no decorative touches, little material imprint of a familial or social world outside. His story is called 'Empty', while his neighbour Mr and Mrs Clarke's story, 'Full', is rich with collections, cooking, home-made exchange, and material reciprocity and generosity among their rich network of friends and family. Miller concludes that his studies 'affirm the centrality of relationships to modern life, and the centrality of material culture to relationships'.

Home is at the core of how people situate themselves in the world and order their past, richly layered in present and future (in other words, the narrative of their lives). More practically, home is a locus for external infrastructure life-supports that are taken for granted in late modernity (sewer, power, telecommunications, broadcast media, water and gas). Importantly for the sometimes legally deviant practice of tinkering, it's seen as a site removed from public scrutiny and surveillance, and by extension a place where some degree of choice and control can be exercised.

Still, there are limits to this choice and control. Stories Tucker told involved professional plumbers and electricians who'd been intermittently called in to help, not because Tucker lacked their skills, but because of regulatory requirements that limit what the home-builder is permitted to undertake. These regulatory impositions exist ostensibly to ensure home-builders reach safe and exacting standards set by various levels of government, but many tinkerers will tell you their workmanship reaches higher standards than accredited tradies. Often, Tucker told me, he had to re-do the dodgy work of tradespeople. In one instance, he noticed the load-bearing support beams in the mezzanine floor had a serious structural error. Two manufactured support structures were designed with a low arch or slight curve that straightens out to strengthen and brace the whole structure when it supports a lateral load. But in this part-housebuild, the support beams appeared to Tucker to be installed upside-down.

So he decided to consult the building draft specifications lodged with council. To his surprise, these files were missing. Nor did the previous owner of the property know their whereabouts. Nor could the original builder find them. This led to much running around and stressful stalling of the project. At one stage an impasse between Tucker and the council threatened. 'Why do you think the drafts were lost?' I asked. Tucker speculated that the files may have gone missing if the tradies who installed the beams had realised their mistake. The beams couldn't be removed without demolishing a side of the house; but a surveyor would never sign off on the job now – as it was, the structure and strength of the mezzanine floor was compromised. If the side of the house was to remain constructed, the project would need tinkering: the two beams would need to be structurally tied together to enable evenly-distributed load-bearing.

Although Tucker had picked up the error, he couldn't legally design this structural tie himself. He was required to engage an engineer to draw up plans. As well as the expense of the engineer, he said, this also cost Tucker a couple of days' work. This story is now inscribed into the fabric of his home. Many tinkerers' stories show how the physical structure of their homes embody the negotiations between tinkering values and those of regulatory and market forces.

When Tucker and I set to work battening the carport, he explained why the Oregon purlins, built by a tradesman, weren't at all suited for a bushfire-resistant home. In the event of a bushfire, 'the Oregon would ignite really easily. It's the kind of wood builders love because it's light and strong, but it's not suited for a bushfire-resistant house.' The battening with cement sheets, then, would protect the purlins from ember attack. The cement sheets had to be spaced so they could expand and contract in changing climate. So we were not to butt them up. With various methods, Tucker had calculated the width for expansion, and the gaps would be filled by an acrylic fireproof sealant.

While it had to be proofed against the smallest airborne ember, the battening couldn't be made completely airtight, because this would lock in condensation between the iron roof and the ceiling. This meant Tucker would have to work out some kind of internal drainage solution. We discussed the properties of various materials, but, as late afternoon approached, the project drifted into coffee-making, discussion of Tucker's curry recipes (a pile of curry cookbooks was stashed in the unfinished kitchen), and the mountain views, wildlife and soundscapes. He played me a recording he'd made of local birds that sang in major keys. It may or may not have been knock-off time: Tucker would see how the evening panned out – he might work into the night. Later in an email, he told me he'd resolved the problem of aerating the battening against condensation by using metal insect screen. He'd also lined the roof with Anticon, a fibreglass-and-foil product that discourages condensation.

On my next visit, Tucker had drawn up plans for what would seem a straightforward project: an upstairs balcony to be built mainly of steel. It would be purpose-built for maximum strength and safety for his young grandchildren. But something was nagging at him when he ordered the steel from the fabricator. He couldn't put his finger on what, but a sense of unease kept waking him in the night. With such a specific custom-build, there was no manual, no-one to advise him, and no tradition he could draw upon. 'I felt isolated,' he later told me. For his balcony to meet regulatory requirements, Tucker engaged a Lilydale engineering company to draw up the first draft. But the engineers hadn't included essential details, and when Tucker requested these, they told him 'that they were accustomed to dealing with professionals'. So Tucker had to calculate the fine details himself. He redrafted, following the building code, which itself had 'minimal' detail in its guidelines. He levelled and concreted the ground beneath the steel, and he worked out how to accommodate its weight, considering what could be braced there, what grade of materials was required here, down to the bolts, the surface treatment, the gauge of balustrade cable and the stainless steel turnbuckles (adjustable wire tension fasteners). The Shire Council surveyor signed off on the plans, not anticipating the trouble ahead.

This balcony would be a big, risky expense for the Tuckers. Facing south, it deviated from the original house-plans, which were consistent with southern hemisphere sustainable building principles that orient all major windows and balconies northward. But an irresistible woodland-valley-scene beckoned in the south, and Tucker had installed double-glazed windows with lovely views onto this aspect. Waking in the night as his unease mounted, he went over the maths and specs with Kerrie. 'You know you're in trouble when you're asking me to do the calculations,' she told me later. Still, between them, everything seemed rightly measured and calculated, and it all had regulatory approval. But then Tucker awoke in the night with a dire realisation: there was a devastating flaw that had gone unremarked by the professionals. He realised he hadn't calculated the steel to allow for the slight drop (25 mm) that directs rainwater down from the elevated balcony floor into the drainage. There was no mention of watershed in the building code, nor on the original engineer's draft. 'You simply are expected, as an owner builder, to know all this stuff,' he told me. Thankfully, this was resolved with a compromise: 'the fabricators cut a little (25 mm) off of the columns – hey presto the required fall to the floor!' Mercifully, the project could tolerate the slight modifications: standard brackets and saddles are designed to allow for a few millimetres' adjustment here and there. And the structure's integrity remained intact. Tucker sealed and tiled the balcony, primed and painted the posts, and ordered the wire, fasteners and guttering.

But by my next visit the project had heaped on more stress. The contractors Tucker engaged to help with the balustrade weren't up to standard, and he spent many hours online and offline in a quest to correct their mistakes. 'The tools the tradesmen were using were clearly not appropriate,' he told me. To ensure optimum safety, he'd chosen a non-standard, high-strength grade of cable-wire that wouldn't give if his grandchildren tried to climb it. The contractors' tools couldn't deal with this grade, but the contractor attempted to push through with inferior tools, compromising the wire's tension and hold. This ended up costing Tucker more time, money, research and frustration. After much running around, he ended up hiring the appropriate tools and redoing the tradesmen's work himself.

Many tinkerers told me that redoing accredited tradesmen's work was routine in their custom housebuilds, which could not be adequately regulated by the standardised approaches of various bureaucracies. Yet as a custom self-builder, Tucker took on all the risk and responsibility, largely in isolation, without adequate insurance, professional support or income, and with considerable personal investment in online research and sleepless nights. In the balcony story, the finished balustrade may have been tinkered for safety and durability beyond professional standards or guidelines, but years into my research Tucker was still negotiating approval of its slight variations with council.

The tinkerer's everyday home life has been largely marginalised from the mainstream story of material production, whose narrative focuses largely on standardised, large-scale mass production *away from* the home. In this story, the industrial revolution saw a shift in production from domestic agriculture (the family farm) to industry; from there post-industrial or information-societies saw a shift from industry (material) to services (post-material). The home in this story, once 'a mere adjunct to the loom or bench', rapidly became regarded as a site of consumption rather than production, according to historian David Edgerton's *The Shock of the Old* (2006). As such, it attained 'the dignified status' away from toil. But it concurrently became feminised and devalued as a legitimate site of productive or vocational activity. It became, according to modernist critic Clement Greenberg, 'a place to leave behind when you start doing something significant'.

Inversely, in the tinkerer's story, the home never stopped being a site of primary significance. It's a place where, according to de Certeau, people tinker (*bricolent*) and thereby enact 'transformations of and within the dominant cultural economy in order to adapt it to their own interests and their own rules'. This is evident within tinkerers' stories, and also by a tally of the types of so-called 'consumables' introduced into the home since industrialisation. Sociologist Elizabeth Shove and design scholar Witold Rybczynski have documented the rapidly increasing demand for home technologies in the past century. Both scholars found that these haven't simply been labour-saving devices such as washing machines, vacuum-cleaners and dishwashers, but productive ones such as sewing machines, power-tools and increasingly 'professional' food appliances and recording devices (photography and audiovisual).

As Edgerton points out, uptake of these technologies isn't solely due to top-down mass-marketing and fashion. Instead, user-generated production and peer-to-peer technology trends are well-documented by scholars. Consumer technologies that have been promoted for specific domestic purposes have been routinely repurposed by everyday users. Technologies marketed for leisure (the car for Sunday drives, the telephone for idle gossip and social arrangements) have become used primarily or to some degree for work; and conversely, those introduced for work (email, web, mobile communications) became important tools through which home-based production and social exchange is shared and promoted. Users – not technology developers – invented the social media conventions we take for granted (such as '@' and '#'). A theory of communities of practice (COPs) – communities whose participants exchange knowledge – has been developed since the early 1990s, challenging orthodox economists' attempts to reduce knowledge economies to explicit or formal exchange of information. The insights of feminist social science, feminist economics and feminist anthropology have drawn attention to the cultural and economic value of unpaid practices of labour in the domestic sphere, and have opened up avenues to understand tinkering's role in economic prosperity. As described in Chapter 2, everyday people, male and female, have throughout history adapted existing technologies (including the family car) in incremental and makeshift ways that became commonplace, commercialised and encouraged by manufacturers until postwar periods.

Technologies in the home are traditionally tallied by economists as 'consumer durables', but Tucker's library of instruments was more accurately described as 'producer investments'. To gather intelligence for his projects, Tucker used an array of precision instruments including a hygrometer (measures the house's relative humidity); a surface thermometer (for documenting the house's slab temperature); a laser-light distance measuring tool to make calculations for building, sound recording equipment, and so on. On his PC he kept digital spreadsheets documenting when and where the house's thermal mass harboured the most heat. These were not the kind of instruments promoted on lifestyle infotainment television or in Bunnings catalogues; they had been collected during Tucker's years of accumulated research among friends and associates, and online. Clearly, this DIY 'passive' home involved a lot of active knowledge, social connection and current technical investment. While the tinkerer might be a parenthetical figure in the labour economy, the tinkerer's home is a site of vocation and useful cultural production, central to life and purpose; not an opt-out from work.

Armed with data from his many instruments, Tucker had dug a trench around the house and insulated it with extruded polystyrene to 'limit the range of temperature variation in the slab through the year'. This would prevent outside ground temperatures conducting into the thermal mass (a concrete slab) inside. He had intimate knowledge of the site and the build requirements, having monitored factors including ground moisture movement and seasonal patterns.

Tucker told me that although he thinks constantly about various elements of the house-build, he often has no fixed plan of what each day holds in store. The shape of his day, he said, is contingent on factors like weather, arrival of components or availability of hire equipment, on outsider tradies, on family commitments, visitors or maybe work commissions, and his mood. He had many projects on the boil, and some days he awoke to see where his inclination and rhythms led him. Occasionally, he'd simply have to wait until momentum hit: with some projects he'd 'go through various phases of incapacitation' before starting.

On my second visit, we'd arranged to talk while installing cabinets in the just-plastered and plumbed kitchen. When I arrived, the day was taking a different direction. Tucker asked if I would instead help him to batten the ceiling of the house's carport entrance, in which he had placed two large aluminum A frame ladders for us to schlepp up the cement sheeting. This was a frequent scenario during my research: on my visit to Gilda Civitico, the arrival of a load of quinces from her parents-in-laws' farm called for a spontaneous change of plan: instead of making jewellery we made quince-jelly. With Adis Hondo (Chapter 7), instead of planing wood we made a frame. And so on.

It's easy to understand these unstructured routines as both a retreat from the impositions of modernity, and as a response to them. Modernity, as Polish sociologist Zygmunt Bauman characterised it, has become in a sense unstructured and shapeless, and as such involves fluxes and shifts in identity, status and security. The advent of late modernity's 'risk society', characterised by casualisation, market liberalisation and deregulation of work, and consequent job insecurity or liquidity, is seen as a force of cultural displacement and of personal loss of coherence and a sense of life-trajectory. If we relocate Bauman's understandings away from contemporary labour markets and apply them instead in the tinkerer's home, these facets of modernity may not be as destabilising or disorienting within the tinkerer's version of freedom. Tucker's story shows that although the tinkerer's home-workplace is indeed a site of risk-taking and uncertainty, and although the tinkerer is indeed subject to all the shifts and perils associated with liquid modernity, the tinkerer's home also holds a library of tools, skills and connections that represent a stabilising (and so thoroughly modern) source of identity, security and agency that might be represented to others by a steady job.

A friend lent Tucker a steel saw-bench, and the previous week he'd bought a second-hand power-saw on eBay to install onto it. But the moment he switched on the saw, 'it arced'. The electrical charge produced a visible flash of current through the air. Impulsively, habitually, Tucker opened it up to see what the problem was. This was a practice from as far back as he could remember. He couldn't remember *not* opening things. Since he was a young boy, he was interested 'in electronics, audio and hi-fi ... Dad trained as an electronics technician and went into the army in the radar scene. After that he had a whitegoods store in Warrnambool.' As well as salvaging parts, 'I also used to rip off parts from Dad's shop.' His father, a sole parent of six children, struggled to support his family, financially and emotionally. 'One Christmas,' laughed Tucker, 'Dad gave me a present: a receipt for all the parts I'd ripped off.'

As a boy in the 1960s he fixed appliances in his father's store. 'And I was always making stuff,' he told me. 'I had a Meccano set. I made a working record player out of it. It had a motor, a wind-up one. You'd wind it up and it had a belt and pulley coming out of it so you could attach a wheel to the belt. In those days Wax Vestas [waterproof matches] came in a tough wooden box. It amplified stuff, so you could hear the record.' Later, when he and Kerrie lived without grid power in their first home, they powered their music by adapting their record-player to the house's wind generator, and Tucker adapted a Honda engine with a generator to operate his welder and other power-hungry tools. He also tooled with Peugeot engines. 'Tinkering with cars,' he told me, 'involved me with a network of other tinkerers'. Both practical information and spare parts 'flowed freely amongst the group'.

Like other tinkerers, he described instances in which his worldview was changed by these DIT exchanges. 'Car repair before that experience was always financial punishment because of the system. It was a test of trust in the unknown mechanic.' In contrast, members of his car-tinkering group 'were trustworthy because of their shared appreciation of the quality of the cars and a commitment to keep them on the road in good condition'. This, he said, changed his relationship with consumer technologies generally. 'Tinkering fed my contempt for consumerism,' he said. Later, he told me he only buys things that can be opened, fixed or modified, and that are 'made of materials that will last'.

But now was the problem of the arcing power-saw. The problem was with the field-coil, a current-carrying component that generates an electric field for continuing circuit in machinery with internal rotating parts. This diagnosis would lead to a multinational quest from Tucker's home. He gave me an abridged version of the protracted story of his online endeavours to source specific parts. He asked me not to mention the brand of the power-saw. It was manufactured by a company that started in the 1970s in Australia, and is now a multinational brand. Like other appliances, some of these power-tools dissuade home-repair by warrantee voids and irreplaceable or not readily available parts. Their components, coming from disparate global locations, often last the nominal time of their warrantees. It's therefore often easier (and cheaper) to discard the whole tool and buy another. In this way, the tool embodies global assembly-line inequalities and discourages tinkering.

But Tucker – like all tinkerers I interviewed – habitually rejects this modern cultural contract with consumption and obsolescence. To replace one of the missing components, he'd visited www.toolspares.com, but was unable to source the component from Australia, so he sourced it from Kingchrome, a UK-based tool wholesaler, which itself had sourced the component from Taiwan, making this project more carbon-heavy than he'd like. He'd negotiated with the eBay seller, who agreed to cover the cost of the component, but only with surface shipping, which would mean a long wait. After that, there was the problem of the saw's customised crimp connector, which you can't buy anywhere. Also, 'you can't solder in there'. So Tucker improvised, making a sheath out of heat-shrink tape. He rescued the saw from a landfill fate.

The repair of everyday objects is not just a material act – it's a moral, social, political and environmental one. It's the fulcrum between so-called First- and Third-World economies: the former dumps its broken and used products along with its toxic waste in the latter, leaving it to the latter to sort and reuse in order to supply First World needs. To Estonian ethnographer Francisco Martínez, acts of repair (or *remont*) 'make late-modern societies more balanced, kind and stronger'. Like Zygmunt Bauman before him, Martínez draws parallels between the displacement, neglect and decay of objects and those of people. His study of Estonian junk hoarders and repairers in a culture of post-socialist transition leads him to believe that 'contemporary mending and the reluctance to dispose of material possession' can be 'a way to resist dispossession and adapt to convoluted changes; the act of throwing away is perceived as a threat to memory, to security, and to historical and ecological preservation'. Product obsolescence, he believes, 'blocks our access to the past'.

One of Martínez's research subjects is Lemmit Kaplinski, who talks to him about the importance of 'preserving industrial heritage'. Kaplinski tells Martínez: 'Innovation stems from understanding old technologies. We can learn from them the basic rules and get past dead end solutions.' From his ethnographies, Martínez grew to understand repair as a form of 're-enactment', of 'healing wounds' and of 'emotional and bodily investment' that bind different generations of humanity together. Technologies are carriers of cultural memory, and repairing them heals us by lending continuity to discontinuity in a fragmented and alienated world. Repair, he writes, 'helps people overcome the negative logic that accompanies the abandonment of things and people, recalibrating synchronicity and a sense of commonality.'

American communications scholar Steven J Jackson, too, believes failing to repair things weakens our connection not just with those things, but with humanity. He writes:

> To care for something (an animal, a child, a sick relative, or a technological system) is to bear and affirm a moral relation to it. For material artifacts, this goes beyond the instrumental or functional relations that usually characterize the attachments between people and things. Care brings the worlds of action and meaning back together, and reconnects the necessary work of maintenance with the forms of attachment that so often (but invisibly, at least to analysts) sustain it.

Jackson believes solidarity with objects affirms our solidarity with humanity, and that the capacity for repair is our measure of a technology innovation's success. The 'efficacy of innovation in the world is limited – until extended, sustained, and completed in repair'.

British craft historian Glenn Adamson has remarked in passing that: 'Perhaps it is a measure of the seductions of capitalism that even its critics tend to think about production entirely in terms of new objects.' (In fact, as other chapters discuss, the global trade in reusable junk is worth billions and is the world's second biggest employer.) In Australia and other post-industrial (and post-scarcity) countries, the tradition of commonplace or everyday repair, reinvention and repurposing has always been acknowledged in popular literature, but has been an 'understudied field' within academic literature, according to sociologists Stephen Graham and Nigel Thrift. Fashion historian Anna König has also observed that the topic of 'mending' is 'an underresearched subject', with 'an issue of low status at play here, one which sweeps the practices of mending from cultural visibility'.

Yet in the past decade, the nascent scholarly field of 'cultures of mending' has emerged. Domestic repair is also an emerging interest of economists, cultural geographers, craft scholars and industrial designers, responding to the past decade's rapid proliferation of open-source repair resources – including community tool libraries, sites such as ifixit.com, repair cafés, Hackerspaces, bike-repair lounges and restarter parties. These are emerging alongside right-to-repair movements and manifestos, including the Maker's Bill of Rights and the MEND*RS Manifesto (pictured in Chapter 2).

The tinkerer's habitual everyday repair could be seen as one of the many micro-routines that amplify to a seemingly haphazard macro one, disrupting or preventing fixed plans. Still, the tinkerer can feel anchored in the particular, regardless of where a project's chain of events leads. And the tinkerer's self-determined approach allows for unpredictability and temporal fluidity that the formal workplace doesn't allow for, and this in turn allows the tinkerer to be at home within a non-committal and indifferent labour culture.

In tinkerers' fluid regimes, uncertainty and waylaid plans don't present insecurity, as they might in a business setting. Instead, they offer more mutable, ductile and resilient work-time cash flow (or resource flow) regimes than offered by a non-tinkerly approach. It's telling that I started studying tinkerers at the height of the GFC, a time when economic growth stalled, unemployment rose by nearly two percentage points, and Australian households' wealth decreased by nearly 10 per cent. Although its impacts were less acute in Australia than in other OECD countries, it's significant that in responses to my questions about work opportunities, risks, income and global trade, not once did the GFC rate a mention, and all but one response to questions about income sustainability were optimistic. Characterised as more flow than flux (Civitico, in Chapter 4, told me her time is 'free-ranging'), more depth than speed (Block, in Chapter 3, said a deep understanding of machinery simply takes committed time), these layered and informal regimes were preferable to formal, linear ones.

In other words, the hassle of a tinkering life is a worthy trade-off for tinkering freedom and fulfilment. Civitico was not alone in saying of her former nine-to-five regime: 'I couldn't go back to that again.' Tucker told me: 'People who become bank managers or accountants miss out on this type of fulfilment.' Many of those I interviewed had never chosen to work in a fixed regime. Adis Hondo, profiled in Chapter 7, said he chose a freelance rather than salaried life because he enjoyed the unpredictability, variety and autonomy. He told me: 'I can tinker around here [at home] doing jobs here and there. I don't have to commit myself to a large time for each part of the process ... It's much better this way.' In Chapter 3, when Chris Block accepts occasional large paid external commissions, he says: 'I want the money so I can *buy back my life*. To buy back my life, you know, so all these other bastards don't have control of my time.' Mark Thomson (Chapter 9), having left his part-time political speechwriting job during this research, doesn't wear a watch: 'My routine is governed by the tension between laziness and the desire to achieve things.' Clearly, the tinkerer's central sense of identity is not with external work but at home.

The values of globalised capital aren't easily avoidable in even the most modest backyard tinkering project, which routinely becomes a multinational, regulated and licensed enterprise, with all the attendant ethical, political and environmental implications. Even within the most ethically-committed workshop, material tinkering can't always avoid using corporate global supply chains, private digital resources, and closed-source materials whose provenance and ethics remain inaccessible to the everyday tinkerer at home. But domestic tinkering nonetheless emerges as good work.

In *Good Work* (1972), the economist E F Schumacher's posthumously-published treatise that spanned economics, sociology, ecology and agriculture, good work is characterised as work that is useful and meaningful. It's work that fulfills our skills, gifts and spiritual needs. Importantly, good work is adaptable to us – not work for which we adapt our personal and social regimes. Good work doesn't work against humanity, but serves and collaborates with it by employing 'human-scale technology'.

More recently, cultural scholars including Susan Luckman, David Hesmondhalgh and Sarah Baker have described modes of creative enterprise as good work, by which they mean mindful work that is meaningful and self-actualising. To sociologist Richard Sennett, good work is a natural result of pursuing the ethics of quality for its own sake, in part because this pursuit provides an antidote to modernity's dehumanising impositions: speed, status-division, industrialisation, standardisation, specialisation, division of labour, mass production, bureaucratisation, secularism and alienation. Against these forces, a 'drive to do good work', writes Sennett, 'can give people a sense of vocation'.

Inversely, anthropologist David Graeber identifies 'bullshit jobs' determined by bureaucratic rather than productive and humanist values; work that generates 'moral and spiritual damage' by employing people in 'tasks they secretly believe do not really need to be performed'. (He nominates these to include telemarketers, PR flacks and private equity CEOs.) The notion of good work is also evident in Crawford's thinking. Crawford sees tinkering as holding profound ethical implications, not only in prolonging product life-cycles, but also in generating self-reliance and individual resistance to modernity's distancing from everyday consumer technologies. This thinking follows a lineage of Marxist and other scholarship that considers agency over productive activity a defining expression of humanity. Marx described this agency as 'form-giving activity' that generates social relations and the 'all-round development of the individual'.

Sociologist Elliot Krause has observed the ways skills since industrialisation have been increasingly organised under capitalist forms (rather than trade guilds and traditional apprentice arrangements) in ways that create a worker whose sense of purpose is withered, with 'the domain of freedom and creativity [reduced] to problems of technique; it creates workers, no matter how skilled, who act as technicians or functionaries'. In concert, Bruno Latour has analysed the ways divisions of labour lead to divisions of knowledge, undermining the ethics of good work. He described the estrangement from material competency that arises from the redistribution of skills from the hands and minds of individual people to machines. In one example, he described how lab technicians no longer need the tacit knowledge, observation and manual dexterity once involved to deploy precise amounts of liquid through a pipette, as a machine is now employed to administer this task. This division of labour and specialisation has seen the skill instead dispersed across a slew of people: designers, engineers, physicists, testers, transporters, packagers, and trades to repair and maintain the machine. A fragmentation of labour and knowledge can distance people from their senses of agency and responsibility. By contrast, good work performed at home reunites these kinds of fragmented knowledges, redistributing knowledge culturally, and reclaiming it personally.

Recently, *bricolage* (tinkering) has been understood by researchers as a defining practice of postcolonial cultures like Australia – a material form of pidgins and créoles fashioned by the colonisers, the colonised and their descendants, and also a method for articulating identity in a globalised world. We have tinkerers to thank for cultural resources ranging from the *iFixit Free Repair Guide to Everything* (as part of the anti-obsolescence movement) to the widespread adoption of the open-source ethos across developed nations. These are material examples of how *bricolage* has for decades been understood as a way of disrupting the syntax of everyday life. It reorganises meaning, and is hence a method of cultural subversion. In Dick Hebdidge's 1979 book, *Subculture: The Meaning of Style*, bricoleurs' transgressions and reorderings are ultimately incorporated into the dominant framework of meanings (commodities, ideologies, art forms) and reproduced as norms, and so there's cultural reform.

In *Renovation Nation*, cultural studies scholar Fiona Allon describes how, as Australians cancelled their travel plans in post-9/11 fear, they 'stayed home and renovated'. So the home became 'a space of control, beyond which potentially threatening and hostile forces played out in random and uncontrollable ways'.

Although Allon presents a thoroughly convincing picture of house-renovation as an insular, bourgeois, reactionary response to the politics of fear generated in the Howard era of border protection and national security alarmism, I came to an inverse understanding of the tinkerer's version of this 'space of control'. To tinkerers, this control originates from a more mindful engagement with the world and with selfhood. The tinkerer's home, located in fluid time, is secured in accumulated skills, knowledge and connections – and the sense of command and confidence that attends all these. It's secured in a comforting habit of temporal rhythms, life-narratives and forebear knowledges intimately bonded with those of other everyday home activities. And its library of tools and technologies represent values that don't fully reflect the bourgeois lifestyle, capital gains and property market motivations that characterise Allon's more fashion-conscious renovators, whose motivations are 'the desire to escape from the world, the felt need for greater levels of comfort and security, or an attempt to exhibit wealth and prosperity, or perhaps all of the above'. Instead, self-imposed home regimes represent simplicity, freedom and sanctuary for the tinkerer.

Still, tinkerers' stories can counter common understandings of *downshifting* and *opting out* for a *simpler life* or *slow life* outside of the rat-race. Susan Luckman offers an alternative characterisation of cultural workers downshifting or tree-changing. She believes that 'a more sophisticated take on place and cultural work would be to say that the creative class, in possession as they are of cultural capital, are seeking to upsize (not downsize) the potential creative capital at their disposal and hence seek places rich in "a spirit of place"'. The tinkerer's everyday home life is in many ways far from a simple cocooned retreat – to the extent that it can seem more punitive, complex and stressful to the outsider than more regulated work–life regimes.

In a 2011 *Utopian Studies* article, ethnographer Nicole Dawkins finds that self-directed DIYers participating in the Detroit Maker Faire or on sites like Etsy are self-exploiting and overworked, and as such are 'replicating the rationalities of neoliberalism that serve to further reinforce structural inequalities'. Among other observations, Dawkins examines home-practice writ public, and she views DIY activity through labour-standard lenses (such as income for hours worked). Given this article, tinkerer Mark Thomson (profiled in Chapter 9), himself a former advocate in the labour rights movement, sent me an email which characterised Dawkins's interpretation as misguided and 'joyless'. He wrote:

> There is criticism [in Dawkins's ethnography] of the 'precarious labor pool' that [DIY] activity involves, which may include – shock! horror! – 'frustration, anxiety and fatigue'. What does she expect? A totally calm (academic?) world in which total order reigns supreme and one day just merges into the next in perfectly formed order and calm? Get a LIFE! Some people – like me – *want* a chaotic life.

John D'Alton (Chapter 8) who also has a prolific commitment to social justice and human rights, offered a similar response, writing in an email that Dawkins's analysis was 'circular'. He wrote:

> Makers make for usually non-economic reasons, and her econocentric Marxist assumptions are showing! ... Because the participants themselves do not feel exploited nor see themselves in that way. They can choose to leave DIY if they want (very very few make a living by it after all). Dawkins has assumed a position and then 'proved' it, by in fact ignoring the actual lived experience of her interviewees.

Marxist scholars might point out that a central feature of worker oppression is that subordinated labourers tend to participate in their own subordination. Yet the tinkerers I interviewed clearly saw the precarity of their work as a worthwhile trade-off for individual freedom that can't be found in formal labour.

To Tucker, as to all protagonists, many of these freedoms are made possible by developments in digital technologies that have extinguished former boundaries between informal and formal life (work/leisure, professional/amateur, private/public, office hours/closing time). Online developments have allowed workplace concerns to intrude more into workers' personal lives, but for the home-tinkerer they have also allowed access to a range of specialisations and communities that suit the domestic autodidact life. In his everyday self-directed routines, Tucker frequently (and at any hour) consults or purchases from online resources local and international – engineering manuals, product specifications, component properties, regulatory bodies, eBay, technology company advice, trade forums and physics information. His online space in the Tuckers' home was established even before the kitchen.

By around my sixth visit to the Tuckers, on a hot afternoon a good year into the build, the property generated the feeling of a home, despite parts still being under construction. This wasn't simply because fittings, trims and tiling, furniture, artwork, a fire-burner and various personal details (fridge-magnet-mounted family photos; a vase of flowers; throws over sofas) were in place, but because the domestic regimes and routines were materially apparent. Vegetables were growing in various repurposed containers on the north-western aspect; washing was hung on a long rope suspended between trees; Tucker's guitar was propped outside of its case against the wall; scattered toys and a pile of dishes bore evidence of a family visit that morning; one of Kerrie's mosaic projects was underway under the balcony; a bookmarked book lay on a table. Kerrie confirmed that certainly it felt like home, rooted as it was in domestic routine. Its material routines (toys would be picked up and scattered afresh; tools used and stored; vegetables harvested and planted; dishes washed and dirtied) had made it a home. Anthropologist Michael Jackson's conviction 'that home is grounded less in a place and more in the activity that occurs in the place' held here, as the activities, habits and stories of generations were manifest. Home lay in material stability, intergenerational continuity, and temporal freedoms. To the tinkerer, the physical house, in all its materialities and temporalities, represents the possibility of home. Home is materially secure but never in stasis. It's never fixed, figuratively nor practically, and this is why, as a routine of continuum, tinkering *itself* figures as home.

To drive home this point, I want to end this chapter with a snapshot of a very different character from Tucker, tinkerer Hulmut Lobb. I first described Lobb in Chapter 1 as a foil to my sample of tinkerers. At the time of my research, Lobb was more or less homeless in terms of accepted standards of physical housing, even though he was a property-owner. Living in Chum Creek, 70 km east of Melbourne, Lobb experienced the very disaster against which Tucker was fortifying his own home. I met him two years after he lost all his tinkered projects – 40 years of passionately collected and assembled radio gadgetry – along with his physical home, in the 2009 Victorian bushfires. 'Nothing was left,' he told me, exhaling. '*Nothing*.' It was 'devastating' to lose a lifetime of accumulated gadgetry. For a while, beset with grief, he felt unmoored, but he regained hope when strangers started donating things to the bushfire relief stock. Old radios, odd circuit-boards, wires, tubes, 1970s telephones, even bakelite vintage radios – the stuff people were passing on 'was amazing', and he felt profound gratitude to people he'd never met but had shown such goodwill. And then Yarra Valley hard rubbish reaped an extravagant trove of parts: antennae, copper, tools, dials. He sourced other bits and pieces online from around the globe, and started tinkering again even before building this, a tiny makeshift shed-home the size of a suburban bathroom, on his burnt-out property. Lobb continued to earn money as a professional pest controller (he was a sole contractor, taking on external work as it came). 'Pest control affords me the time to do this,' he told me, nodding at the components surrounding him. His domestic gadgeteering, in other words, was his primary work-focus.

Paid for by insurance and built according to new fire regulations, Lobb's new fire-resistant home was being built 10 metres away by professionals at the site of the original house. He had hardly to lift a finger towards building it; his tinkering priorities for now lay elsewhere. As he awaited the completion of this bricks-and-mortar home, it was obvious from his makeshift home's interior priorities that Lobb lived and breathed radio-waves. On a desk was a donated computer; on its screen was a radio online forum; on the floor were large piles of donated goods he and his wife planned to put into their new house (quilts, books, kitchenware). But the piles remained messy and unsorted: the shed-home was instead organised around Lobb's tinkering. Lining the walls were shelves: on these an ordered array of valves, tubes, dials, cases and components, on a desk were copper wires, clamps, pliers and manuals. For all its newly-acquired components, this was a scene of routine and continuum, and while Lobb tinkered and talked, telling me stories about his life, he reached habitually for components without glancing at them, and he was evidently entirely at home.

6

Vocation

If her social media fan-base was anything to go by, Kate O'Brien was a celebrity. But she didn't reckon so. Her narrative portraiture – referencing genres from 50s noir and Romanticism to Pre-Raphaelite and Dutch Realism – featured in private collections, public exhibitions, and high-end international magazines, but she later told me these media 'interviewed' her by email. None spoke with her or visited, and nor did she have any idea of what foreign-language magazines published about her. She had not met any of her online admirers. She didn't mind. She told me she was a bit of a recluse, not entirely comfortable with outsider gazes. When I emailed her to confirm my visit, she replied: 'I feel quite nervous about it. My art practice isn't especially linear and I feel a little compelled to run it like a nine to five job for you, when in reality I just … well, tinker I guess.'

At the time of my research, her work featured in a nationally touring steampunk exhibition. You can't really survey tinkering culture without doffing your cap at steampunk. Over the past decades, media commentators have variously declared the steampunk scene officially 'over', or else 'nascent'. (As late as December 2014, the *Huffington Post* was declaring it 'totally now'.) This seems apposite for a culture that frequently describes itself as retrofuturist and neo-Victorian. Steampunk is a subculture that celebrates amateurism in its 19th-century, gentlemanly and dilettantish sense. At its shallowest, steampunk can be empty posturing; at its cleverest, it's a method to criticise the present through lenses of the past. It's an industrial devolution of sorts, but it has fictional roots. The culture

Kate O'Brien's brass plate at the entrance to her home.

is thought to have first taken shape in early 1990s science fiction, in defiance of (or as a pre-history to) cyberpunk's dystopian futurism. But while cyberpunk was concerned with software hacking, steampunk is about hardware hacking – a more material and tactile kind of punk.

In Australia, steampunk is predominantly a Melbourne scene, dominated by an aesthetic that tends to celebrate the Victorian era in a romantic and unproblematic gaze (see Chapter 8). So steampunk'd objects are often contemporary technologies reimagined as if they were invented in the 19th century. A steampunk tinkerer might invent a clockwork laptop or a punch-card brass mobile phone. In turn, the steampunk'd object becomes an argument with modernity – its baroque fussiness is a site of slow contemplation and utopian possibility. It invites a revisionist view of history – a reimagining of technology as a heroic, human endeavour rather than a centralised, standardised, top-down industrial one. *Make* editor Mark Frauenfelder believes it's about the everyman artisan: 'The Victorian era was the great age of the amateur, where nonprofessionals could contribute to the advancement of science, and because these amateurs were most often well-heeled gentlemen, great emphasis was placed on ornamental beauty in their equipment.' *Steampunk* magazine declares: 'We laugh at experts and consult moth-eaten tomes of forgotten possibilities.'

When I met O'Brien, she'd just turned 30, and she lived in a 1970s brown brick house, rented from relatives, in Brisbane's unkempt exurbs, with two petulant cats and her then-husband, who trained in film but worked as a public servant. Much of O'Brien's tinkering, she told me, was supported from his income. Funding for materials, lighting, sets, props and photographic production came 'out of the grocery budget ... when I have an idea, I spend less on groceries'. She earned occasional income from sales or commissioned shots. When I walked up O'Brien's concrete driveway, a stand of scored cat-scratch towers were visible in the front window. Greeting me, O'Brien wore head-to-toe black, beads and crucifixes and Freemason badges (bought on eBay). Her stark black and white hair sat at odds with her face, which I'd seen online in a blonde wig, surrounded by flowers in saturate colours. She told me the black theme was just how she happened to dress today; she *was* working on a series of coffin shots for a group exhibition at a funeral parlour, but these were 'peaceful, not vampiric' – she hated the 'creepy goth girl thing'.

We sat in her dim den doing traditional woman things – me, breastfeeding my newborn in a rococo armchair, and O'Brien settling in a black Chesterfield sofa to continue her needlework with moire ribbons. Her warmth, quick wit and generous conversation invited easy rapport. While we chatted, a hairless grey cat eyed and circled us, Golum-like. The house was more theatre-set than domestic scene. Against the crimson Victorian wallpaper were taxidermied animals, religious and political iconography, a flat-screen television perched between art and history books. An antique typewriter, blanketed in incense dust, was overrun by Freemason paraphernalia and war cockades. And there were feathers, plastic farm animals and baroque picture frames she'd picked up from a Chinese dealer.

Behind the built-in bar lurked a second peevish cat – the house troglodyte. Other rooms in the house were populated with mannequins, busts, hat-stands and hoards of fabrics, antique embroideries, wigs and feathers, and photographic lighting equipment. While we sat doing our business, we chatted about her photographic documentation of Brisbane's ornate Masonic Lodge; her marriage (she met her husband through the online notice-board of 'an obscure New York band'; they married in her early 20s); her rebelling against a Jehovah's Witness upbringing and her 'self-esteem issues' in adulthood. The latter, she said, had given her a Marxist perspective. Groucho, that is: 'I wouldn't want to belong to a club that'd have me as a member,' she said. 'Perhaps that's why I'm into Judaism and Freemasons.'

By many measures, O'Brien was a professional. She was tertiary-trained in fine art. Her photography featured in journals and collections. She paid industry rates to her life-models 'because I don't feel it's ethical for me to ask for their time and vulnerability without paying them'. And, decidedly unlike other tinkerers, she spoke in professional idioms (referring to her 'practice'). But by many other measures she was an amateur, a label with which she was happy to identify. O'Brien explained that although she didn't engage much with the steampunk scene, and she hadn't met Brisbane's steampunk community offline, she shared the steampunk resolve to revive and celebrate a noble 'for-the-love-of' version of amateurism. Her affection for steampunk, she explained, was largely about the amateur way we look at the world and make art; about a heroic, Victorian sense of awe and discovery and possibility – botanical, mechanical, intellectual, artistic and scientific.

Against *tinkering*'s underclass origins, *amateur*'s origins lie in the aristocracy. With industrialisation, the rise of the middle classes and the professionalisation of amateur pursuits, *amateur* lost its aristocratic status. Meanwhile, 'work' during 'leisure' time became viewed as 'a precapitalist vestige of the past awaiting incorporation into capitalism,' according to economist Colin C Williams. After industrialisation, leisure activities became viewed as by-products of the surplus economy, while *amateur* rapidly lost cultural approval with the exile of paid labour from the home. Instead, the word began to function as a rhetorical device to devalue the home as a site of useful cultural production. (Hence, terms of contempt such as 'Sunday painter'.)

The amateur–professional divide became increasingly distinct as the home became gendered (as the back shed or workshop was inversely gendered). Amateurs became feminised figures in Australian culture. Australian essayist Robert Dessaix limns the amateur as a figure resisting the stereotypical male 'onward-thrusting trajectory to their lives, thinking less about direction than about the pleasures of where they happen to find themselves'. By early last century, though, the amateur and the hobbyist, with their 'non-productive' time, were incorporated into capitalist expansion by what became 'leisure industries', including DIY. So as much as the amateur figure was disdained, it also became a valued contributor to economic activity.

By other measures, O'Brien's work also counted as hacking. In the Finnish philosopher Pekka Himanen's *The Hacker Ethic* (2001), the hacker has a committed and prolific work ethic that isn't motivated by duty, money or 'professional' concerns. Instead, hackers are motivated by playful curiosity, exploration, freedom of expression, and passion. Hacking resists supervision and standardised work benchmarks. It approaches work instead as a way of life. According to Australian cultural scholar McKenzie Wark, the hacker 'is a figure that finds its own time' and 'sets its own goals' within a model of 'free, self-organized labour'. Hackers are also largely self-taught, or they belong to communities of knowledge (or what scholars call COPs – communities of practice).

O'Brien explained that she's self-schooled in most of her productive pursuits: studio photography as well as customary techniques of drapery, wig-making, dressmaking, embroidery, upholstering, millinery, paper-folding, soldering, ribboning, painting and jewellery-making. 'I can think of something and then I can find out about it online … I never go, "this is for professionals" and give up. You don't want to be dependent.' During the process, she accumulates knowledge tacitly, culturally and practically. 'When I was soldering, I wondered why something didn't work, and then I figured out I would have to pickle it.' (*Pickling* is a term in metalwork to describe the cleaning of a metal surface of baked-on scales of flux and oxide.) In order to understand her favourite vintage textile site, she was self-schooled in French. 'Ninety per cent of the things online I buy are French.'

She described her skill-set as '*not* professional', but as 'a jack of all trades'. Many gadget-tinkerers also used the latter phrase when discussing their skillsets, but textile tinkering and technology tinkering are differently valued and gendered. Had O'Brien developed the equivalent skills as a technician in as many disparate fields (say,

Kate O'Brien's *Vessel*, 2013.

robotics, mechanics and software), her tinkering might be valued differently.

Domestic tools, methods and materials are all implicated in the gendering of production. A 2014 study published in *Mind, Culture, and Activity* found (unsurprisingly) that a heavy male-dominance is especially apparent in electronics, engineering and robotics projects, while 'crafting, sewing, and other textile design communities attract disproportionate numbers of girls and women'. These disparities weren't simply explained by more obvious social and cultural conditioning, but by social constructions embedded in tools and materials themselves. When the researchers replaced traditional circuitry toolkits with circuitry made up of needles, fabric, and conductive thread, this 'ruptured traditional gender scripts around electronics and computing', leading to the females in their study 'completing highly complex electronics projects by engaging in practices historically embedded within communities of practice with gendered histories'. In other words, women tended to perform well in traditionally-male DIY activities when they worked with the materials to which they were accustomed.

And by disrupting these materials, tinkerers also disrupt the gendering of DIY practices. Kathleen Franz's scholarship on the American automobile history also shows that certain cultural conditions give women freedom and licence to tinker in traditionally male domains. In the Second World War, for example, 'Women drivers tinkered with the car, and by extension, their gender roles.' During that time, she finds, 'motor heroines' who were 'knowledgeable and often independent' provided 'valuable, if short-lived, role models for a new generation of women drivers'.

Despite the current resurgence of domestic crafting, it remains devalued in late modernity. This is in part because it's viewed as rote: unoriginal, repetitive, indistinctive and lacking individuality or creativity. In *The Craftsman* (2008), Sennett describes why home- or folk-craft doesn't hold art's status and authority except when framed in market terms as high fashion. Since industrialisation, art became understood as an intellectual and autonomous practice; and craft as a technical and community one. Art became understood as individual genius; craft as collective replication. More, the virtuoso admired for superior technical ability was superseded by the artist with market and curatorial endorsement. Glenn Adamson has documented the ways the art-versus-industry, art-versus-craft and art-versus-commerce binaries became normalised in the late 18th century, by which time the artist had become a heroic and spiritual figure – in stark contrast to other makers, who carried the lower status of servile manufacturers for mere utility.

But this hasn't always been the case, and many scholars offer alternative ways to conceive of craft, art, and tinkering. Through their scholarship, we can understand *making* as a broad, integrated and multidisciplinary (or pre-disciplinary) practice. For example, art historian John Boardman documents the ways 'art' was once a term that included utility. Art, he asserts, 'had a function and artists were suppliers of commodities on a par with shoemakers' before the advent of museums, collections and art markets. Even then, notes historian Larry Shiner, many non-European cultures continued to have no collective noun for 'art' or 'aesthetic'. Modern polarities of art-versus-craft, artist-versus-artisan, and aesthetics-versus-function weren't fully integrated into European ideas until the mid-17th century, after which 'art' began to be elevated as a discrete system of ideals and institutions (museums, copyright, canons, markets and collections). Through these various temporal and ideological lenses, we can see things that have now become obscure, and we can lean towards some understanding of tinkering's essential breadth. We can see how current views of home-based craft are socially and historically constructed. The recent Western category of *art*, Pierre Bourdieu asserted, is institutionally defined, inseparable from market concerns and deeply linked to status creation.

These marketised systems of value give us limited scope to understand home-based productions like O'Brien's, which are both art and craft, professional and amateur, labour and leisure, love and money. Cultural studies scholars Jack Bratich and Heidi Brush believe the current resurgence of domestic crafting 'has strong links to the anarchist milieu', precisely because of its distance from these marketised values – craft has, after all, relied on hidden (domestic) knowledge and resourcefulness. Domestic craft practice has 'withstood capitalism's founding violence' by being defined by other forms of value; it has endured 'despite capitalism and patriarchy'. And new media is transforming contemporary domestic spaces into public and international spaces. This profoundly changes the conditions of production and the gender values ascribed to them.

Despite various turns of fashion cycles, the status-separation of practical craft and high art, with the latter held in higher esteem since Renaissance Europe, broadly aligns with the status of the amateur. Amateur, traditional and bourgeois home crafts (or worse, 'hobbies') became defined against externalised and bohemianised 'studio craft', made distinct and superior as the link between craft knowledge and fine arts, once inseparable, started to break. To an increasingly official gaze, craft skills performed at home were seen as self-gratifying and therefore inherently unworldly, conservative, nostalgic and uncritical. Adamson writes of a cultural imputation of amateur home crafts in Western culture as almost shameful and self-indulgent. 'So, while amateurism can be the very definition of unconscious cultural practice, it can also prompt anxieties of the most self-conscious kind.'

O'Brien described an incident in which her art was handled by curators of an exhibition featuring her work. Objects she'd crafted from cheap or foraged materials and left to gather dust in her home were then handled reverently with white gloves by gallery professionals. The professionals delicately placed her objects in Perspex cases on gallery plinths. A tinkered prop she'd left around the house was accorded another status altogether once outside her home, in this institutional space. She told me she found this 'hilarious because it had been kicking around the house for so long. I thought, *come on*, I bought that strip of leather from Bunnings and another piece from a two-dollar shop.' She also found herself 'hesitant to put myself in situations where they [curators and critics] are going to analyse my work.' And nor was she comfortable with professional artists' statements. 'I have difficulty explaining what I do in academic language that alienates,' she told me. When a professional artist's statement is requested of her, 'I cobble something together' by drawing upon things others have written about her in magazines. It's not theory, she said, but the technical enterprise itself that brings out knowledge through its own process of discovery. Her visual stories aren't always conceptualised: instead, their narrative and aesthetic trajectories unfold during the material *process* of tinkering. (Since the late 1960s, curators have tended to call this approach *process art*.)

During the shooting of the image pictured, which she says 'is about how after the industrial revolution we gave ourselves up as slaves to the machine', the shutter clicked as the model blinked, and O'Brien knew immediately that this happy accident nailed the story. In the hundreds of proofs in which the model's eyes were

Kate O'Brien's proofs for *Vessel*, 2013.

open, it's clear the compositional, thematic and aesthetic elements are there, but the ethereality, euphoria, innocence and menticide that distinguished the chosen portrait's story aren't present. A slow, uncalculated, experimental and open-ended approach (tinkering) can produce outcomes unlikely to be produced efficiently, linearly or more professionally. And so, once again, the material becomes inseparable from the immaterial, or metaphysical. Rather than executing a pre-conceived idea, O'Brien thinks and feels *through* tinkering.

O'Brien's extravagant process of free-range fossicking, conceiving, crafting and overseeing every stage of production is the polar opposite of a factory model of efficiency, specialisation and management. This model characterises many commercial design studios, and even some contemporary art practice. Increasingly in the past six decades, many of Australia's blockbuster exhibitions feature contemporary artists who direct rather than craft their own work. In *What's Wrong With Contemporary Art* (2004), Australian art critic Peter Timms describes an increasingly industrialised approach to visual art in late modernity. Since the 1950s, arts schools started replacing craft knowledge with art theory, as abstract expressionism, minimalism, conceptualism and pop art became fashionable. Increasingly, art became evaluated and analysed on its conceptual merits over its crafted or technical accomplishment. Timms offers an example: the renowned artist Patricia Piccinini. Her impressive artworks employ a process in which the 'conceptual complexity is not brought into full realisation by the artist, since she doesn't make her artworks herself' – but instead she 'employs a team of technicians and fabricators'. Piccinini, he writes, 'shows no interest at all in process. Having come up with a concept, she has it brought to fruition by others. Her relationship to her art is much like that of [...] engineers and planners.' Timms believes this type of artist occupies the position akin to a distanced industrial designer, because the artworks are not 'the natural outcome of a process', and 'not part of any internal narrative of creation ... the way in which these creatures are fabricated does not contribute to their meaning'.

Against this, as a process-based approach to accumulating knowledge, tinkering serves as a form of scholarship. In O'Brien's process, she becomes a historian. Crafting traditional forms demands a *bodily* and *immersive* knowledge of history, rather than a more propositional knowledge gleaned from books and courses. Her own visual histories are made with old and new techniques of costume- and prop-making, as well as methods of set-building, gleaned from online sources, trial-and-error, old craft books, and adapting or reinventing in her own crafting. One of the thrills of making this way 'is that someone before me has figured out how to do this. I have to try to learn what they learned.'

Cultural scholar Jonathan Sterne sees *techne* similarly experienced – as a social phenomenon that's individually expressed but is 'rooted in learned, embodied social tendencies'. By embodiment, Sterne means the human body, the social body and the material constructions that direct and reproduce cultural practices. To O'Brien, 'photography is just the end of the process of learning how to do things, figuring out things, like how to make a necklace conform to the shape of a neck, why things work and don't work'. In 2009, preparing for the Euchronia (steampunk) ball, O'Brien said: 'Retracing the same stitches as artisans and embroiderers hundreds of years earlier allows me to feel a deeper connection to the history of women and their role in the decorative arts.'

Attuned to the aesthetics and methods of historical moments through crafting, she can plot her projects within larger temporal and thematic narratives. In one conversation, she mentioned her irritation with Baz Luhrmann's *The Great Gatsby*. The hats worn in the film's trailer, said O'Brien, wouldn't be in fashion until 1927, five years after *Gatsby* was set. Nonetheless, she wasn't a purist in her own historical referencing:

> My photos aren't historically accurate … I try to use colour palettes that reflect the time period I'm trying to depict. I'd never go so far to say I have synaesthesia, but I *feel* colours strongly, and think different palettes belong to different times … For instance, I think my thoughts about the '30s come down to limitations in printing methods and age. The ephemera still available was generally printed on poorly calibrated presses and due to age is often foxed and stained.

She told *Modern Day Saints* (online site) that her photographs 'capture a censored and sterilised snapshot of the past … Life's not all tea with Marie Antoinette and Hookahs with Mata Hari!'

This amateur approach and its positivist revision of history are hallmarks of steampunk sensibilities. *Real* in O'Brien's tinkering equates with *material* and *made*, rather than fidelity to accepted or authoritative versions of history. An example of this emerged when I asked O'Brien about her professional competence with Photoshop effects. Much digital skill seems apparent in her photographic portraits, so I complimented her on her skills with Photoshop's special effects. I told her that the steam and Victorian industrial backdrops in one portrait looked very real, even on close inspection. 'They *are* [real],' she replied, adding that she doesn't use Photoshop to doctor her images (she does use the program's filters to make her images flat and painterly). 'My stuff is real: there are no special effects. It might be a constructed reality, but everything is real.' Current technologies like Photoshop filters are simply used to hold the narrative together in time and coherence, making the photographs 'about memories that I create. It's difficult to explain, but they aren't my memories. I create memories for unknown people – the characters I create in my photographs. I guess I get a lot of satisfaction out of exposing a little world my viewers can peek into for a person that never really was.'

Although O'Brien used high-end photographic equipment to finalise a project, she refused to think of herself as a professional photographer. 'I'm not a photographer. It's just the best way for me to package my work so other people understand it.' Reducing her labour to one specialisation wouldn't allow her to tinker. She would 'pretty much spend every day of my life researching and creating wonderful things with complete autonomy', but working this way 'it's very difficult to find regular commissions which allow me to realise the visions in my mind's eye.' While she could do with the money, the linear demands of professional routines wouldn't allow for this type of production. It often took O'Brien weeks, months or years to orchestrate a shoot, as she single-handedly made props, built sets, sewed and sourced costumes, scoured op-shops, purchased materials online, designed lighting and generally had absolute authorship over production. 'I usually spend much more on [a portrait] than I make from it.' She had no project timelines, but many ideas at once on the go. These

A costume and a shoe in preparation.

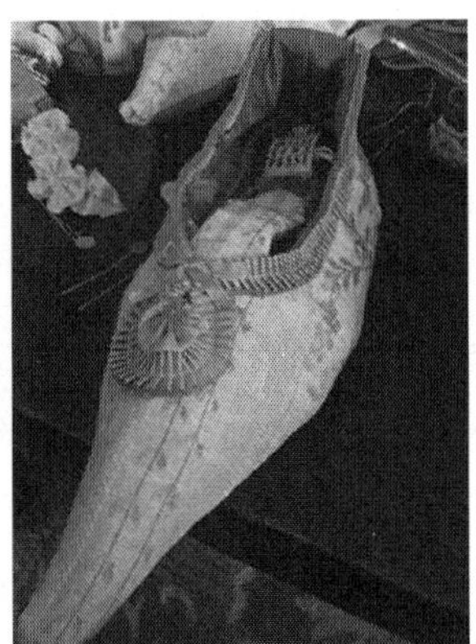

might find their own time for resolution. Or else – like many tinkerers' backyard projects – they might lay for years unresolved.

We talked about a seemingly professional commission: her portrait that graced a winter *FIEND* magazine cover. The steampunk'd woman on the cover wore corsetry, leather, lace, tweed, copper, a fob-watch and goggles. For the set, in warm and green coppers, leathery tans and smoky silvers, she used a steam machine, and, when she took me to a large room cluttered with theatrical props from every imaginable era, I recognised the wall of coppery pipes and bolts she'd hand-built for this single shot. 'I really enjoy being able to build a whole shoot out of my head and make it a reality. That's probably the most satisfying part [but] most of my sets get dismantled or painted over soon after a shoot is finished.' This was only 'due to space constraints'. Some of her images, she said:

> sit in the back of my mind while I collect bits and pieces for them while others only take a couple of months. Many of the props in my images are from chance encounters and there's very little I can just decide I need and go and purchase. Even things like fabric, which is usually readily available, needs to be sewn into a finished garment or backdrop. It's very labour intensive.

In contemporary creative labour like O'Brien's, it's well understood that professionals are not paid according to the time spent on projects. (See, for example, David Throsby and Virginia Hollister's 2003 study, *Don't Give Up Your Day Job: An Economic Study of Professional Artists in Australia.*) The odds of earning a living wage are raised by engaging 'professionally' with the grant and gallery systems. Within these, 'a successful few enjoy considerable benefits in terms of financial reward and recognition, in ways that distort the minor differences in talent that might lead some to succeed more than others,' according to cultural scholars David Hesmondhalgh and Sarah Baker. Without engaging in this system, O'Brien might be considered unprofessional. When I asked her about participating in the grant system, she told me: 'I can't stand the posturing that has to be done. If the work is good, good; and if it sucks, fair enough. But don't ask me to brown-nose my way to the top. I just don't have it in me.'

If you measure her income-from-output, she emerges as thoroughly unprofessional. She might spend months diamond-buttoning, draping and upholstering, and hundreds of dollars and hours on a single shot that never earns a cent. Other shots may be worked upon for years and not ever eventuate. She spoke of her 'guilt' and of her family's attitude in which 'if you can't make money, you probably shouldn't do it, because it's a waste of your life. The general feeling around me is it's self-indulgent, a waste of time … they think I dick around all day.' In *Creative Labour: Media Work in Three Cultural Industries* (2011), Hesmondhalgh and Baker found this a common outsider perspective. 'Although many people respect art, learning and knowledge,' they write, 'some see them as mere ornamentation, and creative labour as a kind of social luxury. This limits the degree to which workers in these industries can feel sure of social respect and recognition.'

Despite her prolific work ethic and commitment to production, in the state or bureaucratic calculus, O'Brien would be counted as 'unemployed' or 'underemployed', terms that imply social uselessness or personal deficit. In *Wasted Lives: Modernity and its Outcasts* (2004), sociologist Bauman identifies an attitude of 'human waste' – a byproduct of globalisation and modernity. One category is refugees and displaced people. A parallel category is 'unemployed' people. Both are outcasts and byproducts of modernity and globalised capital. In these societies, full employment is regarded 'not just as a desirable and attainable social condition, but also its own ultimate destination.' These societies regard 'employment as a key – the key – to the resolution of the issues of, simultaneously, socially acceptable personal identity, secure social position, individual and collective survival, social order and systemic reproduction.' In market economies like our own, people are 'made redundant', as if we were a disposable, obsolescent product, and this has fuelled our senses of fear and vulnerability, making us susceptible to populist politics of inclusion and exclusion.

Our worth is measured by our inclusion in labour markets, but the monetary valuing of work has lost its relationship with work's social and personal value (if this relationship was ever balanced). At a time of rapidly rising economic inequality and wage disparity, historian James Livingston and anthropologist David Graeber both assert that the social and personal value of work has an inverse relationship with its financial rewards. In *No More Work: Why Full Employment is a Bad Idea* (2016), Livingston asserts that our conflation of paid work and good character is 'spectacularly false'. He writes: 'These beliefs are no longer plausible. In fact, they've become ridiculous, because there's not enough work to go around, and what there is won't pay the bills – unless, of course, you've landed a job as a drug dealer or a Wall Street banker, becoming a gangster either way.' In our marketised cultures, we're dangerously conditioned to believe that work is an index of our value to society, he argues. We expect that the labour market allocates incomes fairly and rationally, but incomes are 'completely out of proportion to production of real value'. So participation in the labor market becomes 'irrational'.

Graeber, too, argues that the highest-paid jobs tend to be the least ethical or socially-useful. 'It's not entirely clear,' he writes, 'how humanity would suffer were all private equity chief executives, lobbyists, public relations researchers, actuaries, telemarketers, bailiffs or legal consultants to similarly vanish. (Many suspect it might markedly improve.)' Most corporate and bureaucratic jobs, he believes, are 'effectively pointless. Huge swathes of people in the Western world spend their entire working lives performing tasks they secretly believe do not really need to be performed. The moral and spiritual damage that comes from this situation is profound. It is a scar across our collective soul.' Graeber believes we're crippled by a myth that paid or formal work 'is a moral value in itself, and that anyone not willing to submit themselves to some kind of intense work discipline for most of their waking hours deserves nothing.' And yet there 'can be no objective measure of social value' of work. Against these ideas are ideas of *good work* (documented in Chapter 5), described as work that's meaningful and collaborates with our humanity, rather than alienates us from it – the kind of work in which O'Brien engages.

In the past decade, social economists including Williams have found that work in 'the non-market realm as a chosen space' within marketised societies is undertaken by people 'to get the pleasure that they cannot find in their market activity'. This, they say, reinforces economists' findings that 'a largely unintended effect of a highly individualized and marketized society has been the intensification of social practices' which 'evade the edicts of exchange value and the logic of the market'. In other words, in capitalist societies, a sector of people are reacting to unfulfilling paid work by choosing unpaid but meaningful work. Households with dual incomes within capitalist societies are more likely to seek nonmarket (or 'amateur') fulfilment than those with single incomes. Tinkerers – at least those with partners or those who have financial resources – choose reduction of income and consumption as a worthy trade-off for control over time, deeper familial relationships, meaningful pursuits and better life quality. You'll rarely hear tinkerers talking about the need for 'work–life balance', as if these two are opposed and one needs to compensate for the other.

O'Brien was prepared to commoditise her labour, but only to decent people. 'I have opportunities where I could sell my stuff, where people ask to buy it,' she said, 'but I don't want [some] people to have my stuff, because I invested so much in it'. This was redolent of Civitico telling me that giving her quince products away 'is an act of love, because it's a pain in the arse to make'. O'Brien also told me: 'It's that whole thing, reverse snobbery. I don't want rich arseholes to like my work … but I'm not in a position to turn down money.'

Her ambivalence might suggest a romantic ideal of purity, sanctified away from the taint of commerce, but I see it as a more nuanced mindset in which love-or-money binaries simply make no sense.

Tinkerers tend to regard even their most difficult labour not as toil, but as creative cultivation of self, life and material environment – a view akin to the original amateur-gentleman. And to tinkerers, the prospect of making and spending money in impersonal economies can *detract* from their sense of purpose, pleasure and authenticity. Time-as-money can be – but isn't always – antithetical to the idea of tinkering. Some tinkerers earn well from their skills and projects, and others derive just some income this way. Some don't seek to engage in formal economies much at all. But distinguishing between those who do and don't (and by doing so invoking 'amateur' or 'professional' status) would misunderstand the paradigm of tinkering, according to tinkerer Mark Thomson (Chapter 9). He told me: 'I don't know if you can call tinkering amateur … it doesn't matter whether it's amateur or professional. It's just a gentle method of inquiry.'

Like *amateur*, *tinkering* and *tinkerer* have emerged as antagonyms: invoking both a productive trajectory of work, innovation and knowledge-production; and a useless leisure pursuit of little (or dangerous) consequence. In *Academic Instincts* (2003), Marjorie Garber offers instances of the ways *amateur* and *professional* categories have been defined and redefined against each other, in constant discursive shifts. Alongside the varying statuses of art and craft, the status of the amateur and the professional have been repositioned according to particular ideological and temporal lenses. And so, she argues, *amateur* is a constructed inferiority or superiority, and *professional* has undergone similar constructions. *Professional* is regarded in some domains as high-status, and in others as suspicious (as in 'career politician' or 'professional protestor'). Amateur has, too (amateur sports are approved; 'Sunday painter' or 'armchair psychologist' are disdained). Neither category, asserts Garber, is a reliable predictor of income, social status or competence.

A persistent conflation of non-waged work with amateurism and hobbyism – and, as an extension, unskilfulness – makes tinkering a perplexing idea for 'serious leisure' researchers. 'Serious leisure', according to sociologist Robert A Stebbins, is the 'systematic pursuit of deep satisfaction through an amateur, hobbyist, or volunteer activity that participants find so substantial and interesting that, in the typical case, they launch themselves on a career centred on acquiring and expressing its special skills, knowledge, and experience'. So, to Stebbins, serious leisure can in fact be a career, a term generally associated with professional pursuits. But, he writes, there remains a 'persistent, dominant public view that real personal worth is measured according to the work people do rather than the leisure they pursue'. In *Serious Leisure* (2008), Stebbins finds that people are attracted to project-based leisure because 'they find irresistible its core activity'. Included under the project-based leisure category is 'dabbling, dilettantism' as well as 'making and tinkering'. He categorises cooking, textile crafts and metal crafts as 'tinkering and making'.

In their prolifically-cited 2004 study *The Pro-Am Revolution: How Enthusiasts Are Changing Our Economy and Society*, researchers Charles Leadbeater and Paul Miller recognise that some activities practised within 'leisure' or 'amateur' timeframes are nonetheless practised to 'professional' standards, and that some practices are neither hobbies nor work (or they're both). Yet instead of rejecting the dialectic opposition of these categories (professional/amateur; work/leisure; waged/unwaged), Leadbeater and Miller devise a third tier, identifying a labour segment they have branded (after competitive sport) 'Pro-Ams': amateurs with the skills of professionals but who don't necessarily wish to professionalise or commodify their commitment toward a particular field. These researchers believe Pro-Ams are 'a new social hybrid' that has emerged in the past 20 years. This seems unlikely. Twenty-five years before their report, Stebbins's *Amateurs: On the Margin between Work and Leisure* (1979) also described amateurs who: 'are neither dabblers who approach the activity with little commitment or seriousness nor professionals who make a living off that activity'. And in 1984, too, philosopher Michel de Certeau described how in everyday life 'the dividing line no longer falls between work and leisure. These two areas of activities flow together.'

In the Pro-Am study, women figure less prominently than men as prolific practitioners of skilled home-based or leisure-time material enterprises. The authors propose this is because women are usually primary childcarers and enjoy much less leisure time than men, a proposition thoroughly backed by a body of Australian evidence, which also shows women are generally much more responsible for household labour and childcare. In his 2004 *International Sociology* study, 'The Myth of Marketization: An Evaluation of the Persistence of No-market Activities in a Market Economy', Colin C Williams found that: 'routine housework, cooking, childcare, shopping and non-routine odd jobs and paid work includes time spent in education ... this might well result in an underestimation of time spent on unpaid work and an overestimate of the time spent engaged in paid work.' Academic and journalist Rebecca Abrams maintains that women find it harder than men to allow themselves licence to play in an amateur way, and Sennett describes the ways women historically have been discouraged from leisure or 'idleness' by the church and other institutions. And the raw materials with which women have been traditionally encouraged to tinker – food, textiles, plants – don't tend to be regarded as the stuff of tinkering. Instead, these are the raw materials of the widely heralded 'craft' and 'slow food' movements, or more vaguely as domestic chores, hobbies or home-making.

Despite the success of her art in some gallery circles, there are many measures by which O'Brien's work could be understood as amateur. Yet this 'amateurism' also serves as its own marketing – the marketing of 'authentic'. O'Brien has been *approached* to exhibit, rather than pushing her wares. As a consequence, online commentary lavishes praise on her as if she were a gold-standard folk discovery of purity in an age of inauthentic mass-marketing and monetisation. After the Tweed Regional Gallery made an unsolicited approach to buy one of her portraits, online commenters wrote: 'There was never any doubt in my mind that you are an artist'; 'I've been a fan of your work for quite some time, and am so glad you are starting to get the recognition – and the wider audience – you deserve!'; and 'You have deserved this recognition for so long (as well as the self-acknowledgement of your status as "artist").' To O'Brien's fans, a sense of authenticity and professionalism was derived from her pure amateur values, away from the perceived taint of commerce.

And there were clear profitable impacts from O'Brien's unwillingness to engage in professional codes. Many online sites used her images without permission, and she was tolerant of this, so long as people credit her. 'You can't be precious about copyright,' she told me, 'because these very infringements can get you the work'. When a Dutch theatre company was preparing for a play about the Marquis de Sade, the company came across one of her photographs illegally reproduced on a blog, and it then found her details online. The company offered her a handsome fee – enhanced by a good exchange rate – to use the image in its publicity and program publications.

Kate O'Brien's *Marquis de Sade*, 2009.

The cost of supplying the digital image was insignificant, and in turn the image now comes up in routine searches for the Marquis de Sade. O'Brien doesn't display her work in commercial image-stock banks because she feels they're devalued there. Instead, she leaves it to unorganised, unsolicited and unmediated online media to informally and unstrategically market her work. Amateur or professional, O'Brien's home-product achieved standards of excellence by working competently with globalised technologies (to source, fabricate, promote and distribute), transnationally-sourced materials and globally mobile capital.

Just as they're indifferent to commodity-consumption, the tinkerers I researched tended to consider commodity production incidental or secondary to the more important goal of meaningful experience.

This experience is individually autonomous but deeply communal with humanity. The products of domestic tinkering, though sometimes sold or commercialised, and though sometimes commissioned and made from purchased materials (or paid helpers), generally aren't primarily intended as commodities. Nor is the process of tinkering primarily geared towards professionalisation. Instead, tinkering operates according to an ethos of amateurism-as-love: an end in itself.

Tinkering is neither work nor leisure, but both. It resists the kind of dialectic pairings that characterise modernity: amateur/professional, labour/leisure, love/money, traditional/progressive, cerebral/corporeal, old/new, public/private, consumption/production, individualist/collective, art/industry, tacit/formal, domestic/public, local/global, studio/factory, work/play, ideal/practical, material/metaphysical. The German philosopher Georg Hegel famously believed that the dialectic oppositions within societies will eventually become integrated and harmonised, either through revolution or cultural negotiation. By borrowing this framework, we can start to make sense of contemporary tinkering. Tinkering is a hybrid; a reintegration of the cultural distinctions we've developed since classic antiquity – since *ars* and *techne*. In other words, we can understand tinkering as a postmodern practice, or else as an atavistic one. Both ways, it's a vocation, a form of holism.

7

Risk

Adis Hondo lived many lives. In the split second between hearing a bomb blasting through the window of his Sarajevo apartment and blacking out, he figured his time had come. But he awoke, intact, the next morning. Years earlier, he'd had a near-miss when an earthquake on the Adriatic coast collapsed his hotel into rubble. During the Bosnian genocide, when friends and family went missing, he refused to take up arms, offering instead to take up his camera, and so a POW camp held him captive. He broke his own forearm by dropping a steel girder on it, a tactic to avoid committing war crimes at the behest of guerilla soldiers. It worked. He survived as a passenger in a helicopter whose nose bounced off high-voltage powerlines. He pulled through in an escape from Sarajevo, when he bribed officials and flew out on a UN aircraft to freedom.

His career was equally cinematic. Before the Yugoslavian civil wars, Hondo worked as a cameraman and editor for the state TV station in Sarajevo, filming current affairs, news and documentaries, eventually freelancing for international media. He was offered asylum in Australia in 1993, and, despite the damage the war had wrought on his body, he got by on his native skills as a labourer, carpenter and painter. Learning basic English, within a couple of years he secured work as a freelance cinematographer and film editor.

By the time I first interviewed him, I'd known Hondo for around 16 years through his partner Suze Houghton, my closest friend since kindergarten. He'd turned 54, and was lean and boyish, with scarred legs, rough hands and a soothing voice. He laughed easily. His English was eccentric, sometimes lapsing into mumble, his Rs

Adis Hondo working in his shed.

rolled and his syntax devoid of articles (he'd skip 'the' and 'a' in his sentences, so they amounted to: 'You drive across bridge. After bridge you turn corner.') For these reasons, some of his quotes I use in this chapter are slightly altered and clarified.

When not working on a film, Hondo often sat on a log in the concrete dairy yard in Yan Yean, a dry flat region an hour north of Melbourne, sipping espresso and planning projects on scraps of paper, drawing blueprints from snapshots on his iPhone. 'Sometimes,' Suze told me, 'I ask him if he's bored, just sitting there'. But Hondo told me he's fulfilled and entertained as he sits of an evening on the farm's periphery, imagining how to resolve a project. Among his recent projects was a corrugated iron shed he constructed to stable a battery of industrial-grade power tools and salvaged materials. The shed was within 10 metres from the brick-and-steel remains of an old dairy and its milking sheds that he'd transformed into a home and editing studio he shared with Suze. The home carried what might be recognised by material scholars as the 'aesthetics of decontextualisation'. Its composting toilet was fashioned from an old cowlick barrel (sawdust and shavings from Hondo's home-industry were used here); old bathtubs and disused water tanks now served as vegetable beds; kitchen cabinets and bespoke furniture were crafted from salvaged floorboards; cut 44-gallon drums served as planter-boxes; salvaged windows from various eras flanked the buildings.

Hondo became renowned among his Australian friends for repairing and sometimes adapting machinery. He made bread, spanakopita and *rakija*, a moonshine brandy with a powerful kick. He exchanged his labour. When in Melbourne, he enjoyed free accommodation from a Fitzroy bed and breakfast in exchange for maintenance work. In exchange for helping friends build their homes and sheds, 'Mustafa helped us pour the concrete for the milking shed-house, and Geoff helped with our verandah on the dairy, and part of the floor. Wew [nickname for William] helped us put our first timber floor into the dairy.'

This was how, over many years, Hondo secured the resources to build, wire and plumb this place – a stack of adjoining rooms, studios and outhouses. Some of his tinkering (*rakija* distilling, house-wiring, house-building) wasn't strictly legal, and most of his materials – windows, flooring, roofing, furniture – were sourced from hard-rubbish and salvage yards. This was one of two homes in which the couple lived and edited films, on farmland owned by a family partnership on Suze's side. The couple's other home, an ex-commission house on a quarter-acre block in Castlemaine, was an insurance of sorts, bought conventionally from savings they'd accrued from filmmaking. The risk of this Yan Yean home, said Hondo, was 'that after doing loads of work and spending lots of time and money on the project, the building may be condemned or ripped down by authorities. Or the owners of the land would want to sell up and all the work would be sold with the farm.' This, he said, 'is an ongoing risk ... but the milking shed will always feel precious to us, and the council-approved [Castlemaine] home will always feel like a commodity, until we've tinkered with it enough to call it home'.

Nor was Hondo's tinkering strictly legal in an IP sense. In our second interview, Hondo retrieved from the property's editing studio an orbiting dolly swivel-wheel assembly, a device that allows a camera to rotate 360 degrees around a subject, or, set another way, itself orbits 360 degrees from the one base. He made this device for *Cunts*, a documentary produced by Suze, which relates how a once-harmless word became 'the foulest swear word in the English language'. For this film, Hondo needed a dolly with which to circle smoothly around splay-legged women. A commercial version of this dolly can be cost-prohibitive – independent filmmakers tend to hire rather than buy them. But Hondo copied his from patented designs he'd uploaded to his laptop and iPhone. Patent applications, a bureaucratic device designed to prevent people from manufacturing consumer technologies, serve as rich resources for tinkerers.

The swivel-wheels on Hondo's dolly were salvaged from old skateboards he found on hard rubbish. The body was professionally cut by a laser outfit he sourced in Melbourne. Hondo drafted the cutting patterns, first painstakingly by hand, then with the help of Adobe software. Hand-drafted patternmaking is a rapidly-disappearing specialty in engineering, involving the ability to conceptualise a 3D object and manually draft it with precision in 2D notation, a skill Hondo learned in high school days in Bosnia. Such projects involve a theoretical knowledge of physics and engineering, translated so they can submit to practical skills such as welding and pivoting. Hondo preferred to first draft these manually, with a pencil, than to use digital 3D tools. He said the manual drafting process gave him 'a better feel' for the project.

When Hondo started on his home-made Steadicam for a trip to Arnhem Land, he knew it would be a risky and unlikely project. To many of us, the risk of outlaying time and money to build something with a probability of failure would hardly be worth taking. But to tinkerers this risk is routine. In Arnhem Land, Hondo would join ANU historian Martin Thomas to film a ceremony of the return of Indigenous Australian bones from the US Smithsonian Institution, where they had been since 1948 in a National Geographic natural history and ethnological collection. The human bones were victims of what Thomas called 'a postwar extravaganza known as the American–Australian Scientific Expedition to Arnhem Land'. The bones had been stolen from various sites, 'fetishised and collected in the name of science'.

Travelling first to Washington to film the official handover, Hondo was presented with a problem. Although Aboriginal elders had given their permission to film, the Smithsonian bureaucrats, most probably feeling the shame of the Institution's drawn-out attempts to keep what Thomas called 'its large transnational collection of human remains', withdrew permission. (In an award-winning *Australian Book Review* essay, Thomas would later describe the gravity of the Smithsonian's and National Geographic Society's complicity: 'Theft is a crime against property, whereas this was a crime against people. The removal of bones is closer to kidnap.') This threatened to scupper the hopes of the many Australians who had lobbied and worked for decades to secure the bones' return, and who had pinned their expectations on footage of the handover for perpetuity. Of one Gunbalanya elder, who didn't have a long life-expectancy, Thomas later wrote: 'He wanted to leave a record for future generations. I guess he knew that the footage would be suppressed during the mourning period and that eventually it would resurface.' (Footage or mention of a deceased person by their life-name is taboo until the mourning period finishes.) So at his hotel room, Hondo whipped up a makeshift spycam that he installed into a shoulder bag. He wore the shoulder bag to the handover, later showing me the covert footage, in which the Arnhem Land visitors and Washington officials, including then-Australian ambassador Kim Beazley, are successfully filmed. Stealth cinematography was a skill Hondo had honed during the war.

Back in Australia, there were more complications. It was a privilege, Hondo said, to have permission to film, but it was also problematic. Such Indigenous ceremonies follow traditional protocols but have no scripted events that can be shared in advance with outsiders or codified in a film schedule. This one would involve unpredictable events, sensitivity, and quickness on your feet, relying on a translator for cues and following various subjects and events around discretely as the ceremony progressed. Thomas later described the ceremony's unpredictability and sensitivity in his *ABR* essay:

> Wamud's son Alfred Nayinggul and other senior men and women were waiting, as were Suze and Adis [Hondo] with microphone and camera … Assuming that privacy would be required for proceedings so sensitive, we had used sheets and tarps to screen off a section of enclosed verandah … As with many things that were apparently pre-decided, Wamud reversed his position as events unfolded … He said that we should start filming just before the bones were about to be wrapped. Overnight, however, Wamud decided that the proceedings should be documented more thoroughly.

A Steadicam in this situation would allow for minimal set-up and intrusion, and it was easily adaptable for changes of plan. Months in advance, Hondo worked out its blueprint by trawling through patented designs and observing similar devices in specialist stores, sometimes photographing them on his iPhone. To refine design ideas, he said, 'in most cases you just look around. Since I'm not making things for sale, it's unlikely anyone would know or care that I've copied elements of their designs.' He pointed out that older knowledges and technologies 'that underpin the designs are in the public domain', such as 'gyroscopes, bearings, springs' and fastening elements such as screws. These elements have always been considered open-source, he said.

The Steadicam project was in completion stages, and Hondo showed me what at first blush looked like the bizarre frontal appendage of an angler fish. He said you can run around with such a Steadicam, and turn quickly, but your cinematography remains stable by dint of a gimbal, a precision component that allows the camera to remain independent of the rotation of its support. This precision component was all the tinkered Steadicam needed. Engineering fabricators can make gimbals, but Hondo told me that only a specialist fitter-and-turner he knew could make one for his customised device at a reasonable (informal) price. But then he learned the fitter-and-turner he'd befriended had died. For all his resourcefulness, Hondo couldn't find anyone else in this trade who could custom-make this precision component at a good price. Hondo didn't believe the skills required were far beyond his grasp (he'd learned fitting and turning as a young adult), but he didn't himself have access to equipment to make one. He abandoned his tinkered Steadicam and bought an expensive and legal one off-the-shelf, for around $5,000.

Adis Hondo with his store-bought Steadycam.

But Hondo didn't see the enterprise as unsuccessful or wasted; he saw it as worthwhile. Although he'd invested in expensive materials and countless hours of design and research, in the end the new one had paid itself off within a year. Even if it hadn't, he said, he valued the time and trouble he took to make the version he'd never complete. 'When something is risky, it's always worth it. Even if you can't make it, you know for the next project how far you should go, or where you should stop.' To all the tinkerers I studied, risk equated with knowledge-production. Risk was investment for the next project, and for others like it.

As skilled generalists, tinkerers routinely reunite fragmented, specialised or atomised knowledges. (They also salvage lost knowledge.) By seeing and overseeing stages of production from conception to fruition, tinkerers experience understanding, agency and ownership over all aspects of production – not simply of the thing itself, but of all such things. Upon gaining the tacit knowledge involved in dry-stonewalling, you might notice the workings of walls; after working with wire, you might have a more attuned appreciation for meshing techniques. And you domain-shift these knowledges; you build within your body a tacit potential for *all* craftsmanship. In *The Craftsman*, Sennett asserts that 'it has proved easier to train a plumber to become a computer programmer than to train a salesperson: the plumber has craft habit and material focus, which serve retraining.'

Gilda Civitico, in Chapter 4, told me risk is 'always instructive'. She said: 'If there's a cock-up, it's mine. You're more likely to understand what went wrong if you made the mistake. Tinkering is experimental so the results of your tinkering may be anything from fabulous to bin-worthy. I enjoy this element of uncertainty because I only ever have to please myself and the process is always instructive.' Mark Thomson (Chapter 9) frequently repeated the engineering mantra: *fail early, fail often*. He told me that among his interviewees for his illustrated books (*Rare Trades*, *Blokes & Sheds*, and *Makers, Breakers & Fixers*) risk is habitually welcomed, because tinkering is understood not simply as a vehicle for skills but also a foil. 'I've heard it time and time again,' he said. 'Failure is an opportunity to learn.'

Hondo told me: 'You never know if what you imagine will turn out, and, if it doesn't, often you destroy the material in the process of trying. Or you blunt your equipment working with inappropriate materials. It comes with a certain liability, but when things do work, they make it all worthwhile.' So within tinkerers' experimental habits, risk is understood as a kind of *investment*. This investment is well understood within engineering traditions. According to science and technology theorist Steven Johnson, invention is 'a persistent accumulation of error'. The history of invention, he asserts, 'has a shadow history … a much longer history of being spectacularly wrong, again and again'.

In fact, against-the-odds risk-taking features as one of tinkering's most productive and thrilling characteristics. It motivates the tinkering adventure. To tinkerers, shortcomings of materials and skills aren't themselves 'failures', because tinkering is by nature adaptive: working around limits in resources – material, skill or knowledge. American futurist Alex Pang described this as follows:

> Tinkering isn't so much a specific set of technical skills: there tends to be a pretty instrumental view of knowledge. You pick up just enough knowledge about electronics, textiles, metals, programming, or paper-folding to figure out how to do what you want. It certainly respects skill, but skills are a means, not an end: mastery isn't the point, as it is for professionals.

So, in the tinkering mindset, risk simply can't result in failure. Though tinkerers are often highly skilled and committed makers, they tend to emphasise that less competent tinkerers can't 'fail', because as a tinkerer you are the perpetual apprentice-journeyman (a pre-industrial and unfortunately gendered term). To Gever Tulley, founder of the US-based Tinkering School and author of *50 Dangerous Things (You Should Let Your Children Do)*, risk of failure is an essential principle of tinkering: 'Success is in the doing, and failures are celebrated and analysed. Problems become puzzles … [you] become at ease with the idea that every step in a project is a step towards sweet success or gleeful calamity.'

While the risk of failure is avoided in contemporary work culture, it has an important place historically in traditional master–apprentice instruction. Risk of flaw or failure is also idealised in craft tradition – even where mastery is a goal. It's embodied in the Japanese tradition (albeit in a more refined sense) as *wabi sabi*, or the beauty of imperfection: an idea echoed in Europe in the oft-cited Kantian quote: 'Out of the crooked timber of humanity, no straight thing was ever made.' The Enlightenment principle of 'learning by doing' embodies 'the process of trial and error [resulting in] following a path from many to fewer errors,' explains Sennett in *The Craftsman*. So the experience of failure in manual enterprise 'will cause people to reason harder, and so improve'. In other words, failing is elemental to material scholarship, and it's also traditionally regarded to have moral value. It's iterated in the idea of *salutary failure*, which appeared in 16th century philosopher Michel de Montaigne's essays, in which God disciplines humanity by showing us our limitations.

According to the Australian art critic Peter Timms, in the 1960s and 1970s, when art classrooms were equipped with over-the-counter clay and glaze products and electric kilns, all of which minimised risk: 'Many older potters, justifiably proud of having built their own kilns, dug their own clay and learned from their own mistakes, greeted such projects with dismay.' In his account of the traditional working habits of celebrated Australian potter Col Levy, Timms describes the thousands upon thousands of rejects Levy stockpiles after experimenting with different glazes, clays and techniques, before selecting a piece for exhibition. Timms writes that Levy's failed or flawed pots 'are more important to him than the successful pots, which will go on to be exhibited and sold, because he can learn more from his mistakes, he says, than he can from his triumphs'. It's apposite, then, that 'pottering' has become a term on a par with 'tinkering'. The potter's idealised approach to risk and failure, involving what other vocations might see as inefficiency or wasted time, characterises the approach described by the tinkerers I interviewed.

Risk is also embodied in the romantic histories of everyday manual enterprise. In 'The Nature of Gothic', John Ruskin's famously passionate 1851 essay against mechanisation and standardisation of manual arts, he urged makers to 'neither let your effort to be shortened for fear of failure, nor … for fear of shame'. Within the crudeness of gothic architectural embellishment, in which the craftsman 'with rough strength and hurried stroke, he smites an uncouth animation out of the rocks', there is dignity rather than shame. 'There is, I repeat, no degradation, no reproach in this, but all dignity and honourableness: and we should err grievously in refusing either to recognize as an essential character … or to admit as a desirable character in that which it may be, this wildness of thought, and roughness of work.'

According to Sennett, the craftsman's workshop has historically been a place where: 'Daring to fail evinces a certain strength; one is willing to test why things don't or do work out, reckon limits on skill one can do nothing about'. Despite – or perhaps because of – their high skills and standards, without exception the tinkerers I studied demonstrated fearlessness when it comes to risk of time spent on 'failures'. Chris Block, profiled in Chapter 4, described his foray into terrazzo-making, involving experimentation with different aggregates and curing. A repeated series of failed prototypes spurred Block on to develop a new method in which he uses a wok for mixing and, within handmade wood frames, he 'sinks' the aggregates onto the flat underside, instead of rolling them from the top, as is the traditional method. He showed me his terrazzo tiles, embedding impressively fine numerals and letters, and when I asked where he sources the confidence even after repeated attempts fail, he told me: 'It just doesn't occur to me that I can't work out how it's done.'

Tinkerers routinely expressed this attitude. Macclesfield tinkerer Adrian Matthews ran all his vehicles on used vegetable oil that he doesn't convert into biofuel, but simply filters through a panty-hose. He told me several people, including mechanics, warned him it was risky to the engine and that it wouldn't work. But in his four vehicles he widened the engine hoses to allow for the oil's viscous properties, and modified them to administer a small shot of mineral oil to start, after which they diverted to the vegetable oil. They've run like this for years, and Matthews told me: 'I just didn't believe it couldn't be done. I didn't believe I couldn't do it.'

This was echoed by Mt Barker-based builder and inventor Phil Gerner, who worked with Mark Thomson on many tinkering projects. When they built a giant toroidal vortex smoke ring generator out of a corrugated-iron water tank, he told me: 'I don't think about it not working. I don't think about not doing it.' Nairne-based tinkerer

Eddie Banks, building his family home, told me: 'You just have to give it a go. It's that thing, just having a go, working it out.' His partner, Nina Keath, added: 'The thing with Ed is that he's not afraid to fuck it up. He's a perfectionist, but if a project fucks up he doesn't get stressed like I do, he just sucks it up and keeps going until it's resolved.' To Mark Frauenfelder, editor of *Make*, this approach is the very nature of tinkering. Frauenfelder came from a background in which he was 'trained to believe mistakes must be avoided [and so] many of us don't want to attempt to make or fix things'. He wasn't a native tinkerer; he was only able to become one after shedding habitual risk-averse behaviours and attitudes. Observing tinkerers, he wrote that 'their secret isn't so much what they have as what they don't have: a fear of failure'. When I put this idea to Hondo, he shrugged. He felt making anything is simply a series of problem-solving steps. 'Problems,' he shrugged, 'are only problems'.

Tinkering stories everywhere carry this problem-solving valour. As my study was in completion stages, the ABC's *Australian Story* featured Nimbin-based tinkerer Cedar Anderson, who'd been tinkering for decades to invent the Flow Hive, a device that broke world crowdfunding records (it raised $12.2 million in a month). His mother told the program: 'When Cedar gets an idea, he's utterly tenacious. He really believes that if the human mind has worked something out, he can too.' In my study, tinkerers' material tenacity and risk-taking was often attributed to tinkering families – families who bestow confidence in their children's' manual enterprises from a very early age. Likewise, in *Australian Story*, Cedar Anderson's friend attests: 'I don't think anyone really ever told him what he had to do or that he had to pursue a certain career. And his upbringing allowed him to, you know, nurture his own natural curiosity and follow his own instincts.'

Anderson told the program that his childhood community in Nimbin encouraged kids to be free and take risks: 'We built this electric shock machine. I can remember it was a hit amongst all of us kids. We'd join hands and sit around in a circle and hold onto these terminals and be getting this ding-ding-ding-ding-ding.' He remembers fondly the electric-shock hazard. To *The Art of Tinkering* authors Karen Wilkinson and Mike Petrick, tinkering involves 'sometimes dangerous situations … the dangerous aspect of tinkering is a powerful motivator'. Among the tinkerers I studied, Eddie Banks's story demonstrates this. His family allowed him to almost single-handedly build an amphitheatre at the family's property in Nairne for a Year 12 project, giving the teenager funds and freedom to hire a bobcat and other equipment to terrace the land and build a hardwood stage (to this day, the amphitheatre continues to be used for performing arts events).

In Hondo's case, among Bosnian families in the village where he grew up, he says it was 'unthinkable' to outsource labour to outside tradespeople when building houses. 'Your family and friends build it. You might get one professional, but everyone around is just helping.' During (Josip Broz) Tito's governance, he said, schoolchildren were routinely taught to make electrical circuitry, drive vehicles, draft designs and engineering patterns, work with wood and metal, and cook. 'Even some might fly aircraft.' In after-school clubs, under the tutelage of a senior professional, they learned electrical engineering, camerawork and aerodynamics. In his very first lesson in metalwork, around the age of nine or 10, Hondo and his classmates were given a piece of steel and a file. They had to make a key to fit a lock. 'After that you move to a complex project, then you go to a factory where a retired professional teaches you. You do basic things and then go from one machine to the next. You have to repeat the year if you don't pass.' Even the most complex technologies didn't seem beyond Hondo's grasp. Later, Suze told me her visits to the Balkans impressed upon her how culturally-embedded his values are. She said when the Socialist Federal Republic of Yugoslavia 'went out on its own, it had to be self-reliant' and she believed its people internalised this ethic. She told me this made the production of high-end export technologies 'like pace-makers and helicopters' feasible for domestic industries.

Estonian ethnographer Francisco Martínez also describes how historically 'a different material attentiveness' developed in parts of socialist Eastern Europe, where 'repair was praised in a context of shortage and mass production, in which planned obsolescence and competition, both essential to a market economy, had no role'. This different attentiveness 'also related to particular skills that are not appreciated in an economic system that relies on accelerated cycles of production-consumption-disposal and rapid financial profit'. Privatisation and competition are often spruiked as encouraging innovation, but Hondo said the values of short-term profit suppress the risk-values of invention and repair. He frequently expressed his disdain for our obsolescent product-systems. He was scathing, too, of the kind of bureaucratic risk-aversion that health and safety regulations emphasise in Australia today. So were Australian-raised tinkerers – especially Mark Thomson, as described in Chapter 9.

Hondo told me it was a shame most Australian kids aren't given the opportunity to acquire the lifetime skills he learned because our schools have become so materially risk-averse. But Australian-raised tinkerers told me this too. Chris Block and John Tucker lamented a loss of manual skills in Australian classrooms; as did Mark Thomson, who worked with Big Picture Education to promote tinkering projects in schools. Years into my research, I noticed many tinkerers had mentioned that our highly-regulated school system – in which children sit silently in rows, clocking on and off within strict regimes to produce standardised outcomes – was designed in the 19th century to train a factory workforce. Thomson was among many who complained that our schools discourage the risk of failure, a central feature of tinkering. The principle of trial-and-error, he said, 'does actually entail some error'. He said: 'Part of it is the classic experimental mindset – nothing ventured nothing gained.' Mister Jalopy, *Make*'s most celebrated tinkerer, concurs that 'the act of failing again and again' is the only way to become a successful tinkerer. Thomson described the risk of failing as 'essential'. He said a fearless 'can-do' attitude is usually passed on through familial role models, and testimonials from Australian tinkerers online seem to support this claim. On the ABC website, for example, 'Kate from Brisbane' reports:

> I come from a family of tinkerers. It started with my grandfather, he was a melting machine mechanic, but it's my Dad I remember. He really helped my sister when her car broke down, and she was out of town. The heater exploded and spluttered, and so she went to a mechanic, he said it would be a week. So she rang my dad and he said, 'Put the mechanic on', and he said to him, 'Now just get a broom handle and stick it in the hose, she doesn't need a heater', and he sent her on her way.

At one stage, my discussion with Hondo moved away from the values of makeshift repair and salvaging and instead focused on the value of a mass-produced petroleum-based commodity product – the plastic chopping board; the coloured kind you see at cheap import shops, otherwise known as '$2 shops', that carry the sour odour of sweat-shop plastic that's toxic to its Third World makers and inevitably ends up in oceans and landfill. In a moment of inspiration, this common object had occurred to Hondo as he sat in the dairy yard. He figured that plastic chopping boards could perform many tricks. They're cheap, and the right profile, and exactly the lightweight, durable, easily whittled material he needed. Hondo whittled a red one into the precisely-carved saddle that now formed the seat on which to secure the tripod of a dual-bar jib-crane he designed for his paid work as a cinematographer. To make the crane operable, he welded and bolted steel bars and pivots.

The project then presented a quest: for the jib to remain stable, Hondo needed to connect heavy axis counterweights distributed evenly on both sides of the arm. He pondered various components, but the serendipity in the jib-arm story arrived when he happened to notice some gym weights gathering dust in a friend's garage. As Hondo recalled it, his friend told him: 'You can have them, they're just in the way.' The gym-weights had a threaded bar through them, so he could hold them in place and adjust them up or down the bar, using end nuts. Hondo envisaged using this original weight-bar for the arms of his camera-jib, but a couple of raised rims on the arm made it unworkable. He resolved this by instead buying some threaded rods and bolts with which to hold the weights.

But then the weights clanked together when the jib panned. This wouldn't do. Noise is a hazard on film sets, so to prevent this, and to allow fine-tuning of the weights' positioning, Hondo used hole-saw bits on his power-drill to cut into more plastic chopping boards, making them into uniform circular disks. The finished jib allowed his cameras to tilt at any angle and pan rapidly but smoothly across locations noiselessly, and without tipping.

Adis Hondo's home-made jib-arm, with circular disks made from plastic chopping boards.

Still, the crane remained risky and illegal 'in the occupational health and safety sense'. Hondo described how an earlier version of the crane 'if mishandled could kill someone or damage them badly'. The only time anyone else used this first incarnation of the crane, the brake on the tripod head stopped working, and it became dangerous to operate. Luckily nobody was hurt. 'But I resurrected it by reducing the size of the arm to a half sized model, and that seems safer to operate and more useful for the kind of films wc make.'

An equivalent commercial version of this crane costs around $6,000. Hondo estimated his one-off version cost a couple of hundred, plus hours spent contemplating in the yard, drawing plans and tinkering with wooden prototypes. Australian film industry colleagues offered him commissions to make this same cheap model for them, but he declined. 'I make for myself,' he said. 'Too much responsibility to make for others.'

Such products are too risky to commercialise, and not simply because they can be hazards. While the products of home tinkering can sometimes be marketed, many are neither commodity forms nor reliant on commodity forms – in part because contemporary consumer products are designed and regulated to actively discourage tinkering (as we saw in Chapter 5). This might suggest tinkering is mindfully oppositional, but Hondo's story invites a more nuanced interpretation. It shows that although tinkerers value thrift and tend to oppose product obsolescence, they're not always (or not necessarily) *against* commodity-consumption, but indifferent to it.

Chris Block's story in Chapter 3 revealed that one of the most thrilling aspects of tinkering is the serendipity of salvaging the right component, or the exhilarating feat of adapting the 'unsuitable' component. If such fortuitous events don't happen during a project, and if less formal exchanges (junkyards, op-shops, online trading or friends) are unfruitful, then a trip to the hardware store is no sin. It's simply a practical strategy. In Block's case, when it became impossible to find one component, he simply paid for a laser-cutter to fabricate it. This is how tinkerers' process-based and opportunistic materialism can be distinguished from the materialism of commodity consumption and production.

So while counter-cultural opposition can be easily identified and co-opted by merchandisers (think, for example, the licensed Che Guevara keyrings and trinkets at thechestore.com), the more risky and opportunistic practice of tinkering makes it a moving target for marketers. Although there have been attempts to capitalise on the risk of tinkering within IT and steampunk realms, and in magazines such as *Handyman* and *Make*, everyday material tinkering is elusive because it operates in an indeterminate niche. Hondo's tinkering, like most tinkering projects I studied, involves a combination of informal and formal solutions, as well as salvaged and consumer solutions. Occurring in the private domain of risk and privacy, tinkering projects' discreteness and specificity tend to entail unorthodox, hybrid and sometimes-illegal DIY solutions. So tinkering can never be co-opted by the formal economy and its values. Tinkered projects often *devalue* commodity forms in formal economic terms, but tinkerers also add layers of personal meaning, currency and memories beyond those that commodity forms can offer.

Risk is the primary difference between a crafted product and a manufactured one. The distinction between craft and manufacture is elegantly theorised by the late furniture designer David Pye, in his seminal *The Nature and Art of Workmanship*. First published in 1968, Pye's theory continues to rank as 'the purest piece of "craft theory" written in the twentieth century', according to craft historian Glenn Adamson. Pye argued that it's erroneous to make artisanal claims by using the misnomer 'handmade' or 'hand-crafted'. Do our products no longer count as 'craft' or 'handmade' if we used a power-saw in their making? Or a lathe? Or a food processor? Or a camera? Pye believed these are the wrong questions. He instead distinguishes between the 'workmanship of risk' and the 'workmanship of certainty'. A manufacturer using precision instruments within a highly regulated system and predictable materials is engaging in the workmanship of certainty, and therefore produces standardised and uniform products with very little scope for error. (Think: a McNugget.) On the other hand, an autonomous worker might use either basic hand-tools or highly specialised machinery, but regardless of which, can still practise the workmanship of risk where the margin for error is much greater than on the highly regulated factory floor. (Think: an artisanal sourdough loaf.) So the workmanship of risk can be understood by degree rather than kind. 'Handmade' and 'handicraft' are historical or social terms, Pye argued – not technical ones. His distinction is useful when considering the difference between hardware and software tinkering – on the continuum between 'risk' and 'certainty', the latter can be plotted more toward the right. Working within standardised codes, software tinkering generally has more predictable immediate outcomes. For the hardware tinkerer, bodily approximation, care, dexterity and aesthetic or technical judgment are all human means of reducing risk in workmanship.

So despite the current trend of 'craft-washing' (Pepsi is spruiking its drinks as 'craft soda'; Domino's marketers came up with 'artisan pizza'; even McDonald's launched an 'artisan grilled chicken'), mass-producers simply can't make 'craft' or 'artisanal' products. These words refer to autonomous human-scale production that's too mindfully- and bodily-involved for the assembly-line. To a craftsperson, conception and physical production are inseparable, and the maker's relationship with their craft – be it breadmaking, songwriting or welding – is somatic and risk-ridden. Division of labour completely wipes 'crafting' from the fabrication process. Manufacturing involves predictable and uniform outcomes; craft involves risk and unpredictability.

Gell's theory of art and magic in Chapter 3 suggests why the highly skilled products of the workmanship of risk – rather than low-skilled products of the workmanship of certainty – are coveted by societies discontented with industrial standardisation (hence the corporate craft-washing campaigns). The products that are highly risky to produce are considered more 'artisanal', and therefore more magic, than those produced with high degrees of certainty – not just for the beholder, but for the maker. From the perspective of the tinkerer, an important difference between workmanship of risk and of certainty is that the latter is, as Pye put it, 'simply by its nature, incapable of freedom'. This is the freedom of unregulated thought and experimentation. The workmanship of risk, in other words, is at the very core of creative freedom, a mode-of-being regarded by the sociologist David Gauntlett as 'one of the most central aspects of being human'. The freedom to tinker is borne from risk.

For the skilled tinkerer, the materials themselves often pose a great risk. This is why tinkering remains a subplot in the story (or marketing) of DIY's resurgence. Since the 1970s, many raw materials (such as hardwood and copper pipe) were domesticated (as MDF, fibreboard and PVC), transforming and extending the kinds of projects the layperson was able to attempt. In contemporary DIY projects, competence is enacted not just by a person's skills or by his or her tools and machines, but by many of the purpose-designed materials he or she may integrate into the project. In *The Design of Everyday Life* (2007), Shove et al. build upon Dant's and Latour's ideas of materials as embodied knowledge, explaining how 'intelligent' paints, or products like Speedfit plumbing, have immutable agency. They 'rearrange the distribution of competence within the entire network of entities that have to be brought together to complete the job at hand'.

This agency is invaluable for the domestic DIYer whose goal is to fulfill the material's *intended* embodied tasks – in other words, if the goal is to perform the (standardised) workmanship of certainty. But materials are mutable in the tinkerer's mind, regardless of their embodied intent. Unlike the DIYer constructing a step-by-step IKEA cabinet, or custom building with more traditional trade materials, the tinkerer's approach very often *revokes* embodied knowledges, repurposing or reassigning materials beyond their designed use. As Chris Block's story showed (Chapter 3), much of tinkering's risk arises when already-made material products are freed from their normative roles and reimagined to perform others.

To anthropologist Arjun Appadurai in *The Social Life of Things* (1986), 'the diversion of commodities from their customary paths always carries a risky and morally ambiguous aura'. Earlier in this chapter I described what material scholars call the 'aesthetics of decontextualisation' that characterised Hondo's home. In some form or another, this decontextualisation was apparent in all the tinkerer's homes I visited. And I liked them: these aesthetics transmitted to me a kind of aura, or magic. If I were to accept outright Gell's argument outlined in Chapter 3, I might consider this aura to be generated from the transformative risk these homes and their objects embody.

But other scholars would see my attraction as hopelessly ignorant and bourgeois, bound up in my own nostalgic, romantic and fetishistic impulses and biases. To Appadurai, the value of made objects in 'highbrow Western homes' is enhanced by placing unlikely things in unlikely contexts, where the objects are diverged from their commodified value and thus bestowed with a decommodified 'authenticity'. To Appadurai and others, this quality of authenticity is also derived from a quest for novelty and a sense of the exotic attributed to the original object (parallel with the Marxist idea of mystification involved in commodity fetishism). To some scholars, this fetishisation can be born of cultural ignorance.

But these scholars' concerns sit largely in the context of First World practices of collection, decoration, and postcolonial souveniring. They're not so much concerned with the objects' new function. To the tinkerer, aesthetic decontextualisation is often born of calculated risk-taking and generated from deep knowledge and deep understanding of material properties and potentials. In Hondo's case, a dairy shed, salvaged floorboards, skateboard wheels, cowlick barrels, chopping-boards, gym-weights, bathtubs and disused tanks serve utilitarian purposes as much as aesthetic ones. In some tinkering projects, such as Michael Drinkwater's conversion of a shower out of a Telecom booth and a biodiesel processor out of a dishwasher (the economics of which are discussed in Chapter 4), decontextualisations are rudely utilitarian and not for aesthetic affect.

Even projects that emphasised decorative concerns as much as utilitarian ones embodied risk. This is the risk of domain-shifts between material ideas that were common to every tinkerer's home environment. Irene Pearce's tank-home was the most dramatic example. In 1998 Pearce, recently divorced and grieving for the loss of two of her children to separate car accidents, spent her modest savings on a block of land in Mount Barker, South Australia, on which a 50,000-gallon concrete tank had stood since 1944. Although at the end of a residential-zoned cul-de-sac with impressive pastoral views, the block was cheap and had been on the market for years because this tank, partly set underground, was so well-engineered that it would be prohibitively costly to demolish. But Pearce regarded the tank's recalcitrance and intractability as assets, rather than as liabilities. Gradually, obsessively, using solely salvaged materials, she tinkered with the tank until it became a home.

Like Hondo and Drinkwater, Pearce made the home's furnishings, fittings, landscaping and decorations by salvaging and reassigning objects' original uses. Where she lacked building skills (her background was teaching sculpture and ceramics), she improvised. When salvaged objects refused to submit to their new use, or when Pearce lacked the skills to craft them, she invented unorthodox ways to will them into compliance. The slabs of ironwood she salvaged for her kitchen benchtop (pictured) didn't meet at the joins,

A 1980s Telecom phone booth that has been repurposed into a shower in the shed-home of Benalla-based tinkerer Michael Drinkwater.

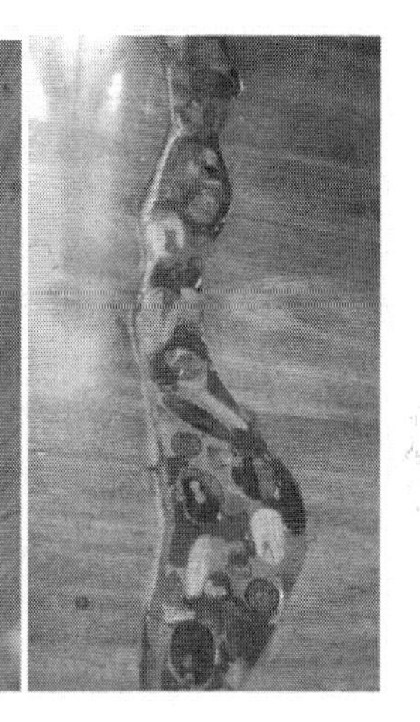

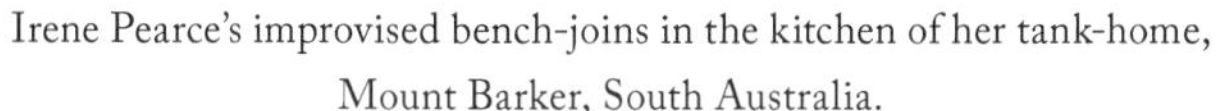

Irene Pearce's improvised bench-joins in the kitchen of her tank-home, Mount Barker, South Australia.

were irregularly shaped, and weren't the precise lengths required to cover the cabinets. Pearce had neither the skills, physical strength nor the equipment to cut them or join them seamlessly, and nor did she have the money to pay a professional. She considered asking friends to cut the ironwood, but an important part of her healing process involved realising her own autonomy and resourcefulness. So, instead, she made a sculptural feature of the project's 'failings'. Where the joins didn't squarely meet, she embedded judiciously-cut stick-profiles and decorative wood scraps from other projects into epoxy and wood-putty to fill the gaps, and varnished over them and the ironwood to make a smooth decorative surface. Only through risk-taking, wrote Pye, do we have the freedom of unregulated thought and experimentation, unknown outcomes, exploratory learning, and idiosyncrasy and diversity rather than conformity. Pearce's irregular joins (pictured) are now treasured features of her kitchen. Rather than hiding their histories within the anonymity of generic or standard bench-top joinery, the bespoke bench-tops

Adis Hondo in his Yan Yean shed. Photo: Greg Rochlin.

embody the story of Pearce's hope and healing, and also of her tinkerly risks and resourcefulness. Tinkering is indeed the embodiment of hope.

Pearce told me: 'I don't see problems as problems.' Hondo would tell me something strikingly similar: 'Problems are only problems.'

Problems are only problems is how Hondo approached life. In any situation, under any stresses, he was unflappable. Before I finished my research, Hondo's many lives ended. He was born 26 April 1959 in Mostar, Bosnia. He died in Castlemaine, Victoria, on 7 December 2015. Many people miss him, in Australia and in the former Yugoslavia. Peerless in his generosity, he made and fixed things everywhere, for everyone. He was a humble and magnificent man. Suze is still living in his material legacy, and in many ways so is he.

He found joy in ordinary things. For twenty years, even in his dying days, Adis Hondo showed me how to see what I was looking at.

8

Utopia

Matter-of-factly, John D'Alton said to me, 'I identify with the Wizard', as if this explained a poster I was gazing at. The poster was hanging in his vast study, but the room wasn't built as a study. It was the basement billiard room of D'Alton's 1970s home in Greensborough, a bushy outer suburb of Melbourne. Its mock-Tudor ceiling and electric copper lanterns emitted the faux medieval atmosphere favoured by Irish-themed pubs. Under the billiard table were computer carcasses and gum-tree twigs – props left over from *Eucalyptus Wars*, a D'Alton-authored LARP (live-action role-playing) game. Edging the walls were stacks of books organised under broadly utopian themes: Anarchist Social Movements, Social Revolutions, Appropriate Technology, Early Islam, Early Judaism, Early Christianity, Sustainable Living, Science Fiction, Mysticism and Speculative Fiction.

The room had a wet-bar but no apparent alcohol: behind the servery was a large inflated *Dr Who* Dalek, its eyestalk and deathray poised to *Exterminate!* Overseeing the Dalek was a poster of the thing's spiritual nemesis, the deeply intellectual *Star Trek* Captain Jean-Luc Picard, master of diplomacy and sage wisdom. Shelves built to hold liquor were repurposed to house aliens and humanoids. There were boxes of ambush games, stacks of *Wired,* and wobbly-headed Thunderbirds; on the bench were an electronic keyboard, sound mixers and synthesisers. Where it wasn't crammed with books, the opposite wall hosted Victoriana including a sampler box of airship knots made from op-shop pickings. Memorialising *Babylon 5* and religious figures alike, the room was a shrine to time and space, allegory and history, war and peace, cultural and creature diversity.

John D'Alton and Connie Chen.

And so there were clocks and cultural and religious calendars. Like his Dr Who muse, the real-life D'Alton was entirely at home time-travelling as a missionary among aliens. Among his roles – tinkerer, Christian chaplain, activist, journal editor, university ethics advisor, role-playing game-master ('us RPG writers are all failed movie script writers') – D'Alton was a historian, completing his doctoral thesis on 'The concept of struggle (jihad) in pre-Islamic Syrian Christian and early Sufi Muslim writings'. He served on the board of the Jewish Christian Muslim Association, and, like Captain Picard, he had a missionary interest in peacekeeping between divergent worlds. 'People are unaware,' he told me, 'of the ways Jihad as an internal battle features in Christianity'.

D'Alton's warmth and optimism were infectious, his proselytising gentle and erudite, and each time we met I left the place feeling as if we'd engaged in some kind of *satsang*. He worked as an Orthodox Christian priest and chaplain at La Trobe University, and he spoke with the soothing upbeat assurance you'd expect from such a person (or role). Sitting or standing, his posture was presidential; he looked 20 years younger than he said he was (51); but a Lenin-style beard also made him look like the kind of person who smokes a pipe and hunts deer. By his desk in this basement were various religious icons – most arrestingly, a small framed portrait of the Russian Orthodox St John of Kronstadt. Its subject looked so remarkably like D'Alton himself that I presumed he'd tinkered with it in Photoshop. He hadn't: D'Alton told me that Kronstadt, famed for his almost superhuman commitment to social justice, is his inspiration. Later, he said the same of Ghandi.

Perhaps sensing that his Wizard comment might demand some explanation, D'Alton took a book off the shelf and waved it at me. It was *King, Warrior, Magician, Lover* by Robert Moore and Douglas Gillette. 'I identify with the *archetype*,' he explained. 'You familiar with Jungian archetypes?' Later, when I returned home, a Google

John of Kronstadt, 1860s (left) and John D'Alton, 2009 (right).

search joined some dots: wizard = magic = mysticism = early religion = man-of-God = carpenter. Or another role-playing pathway: wizard = magician = transformation = tinkerer.

Bizarre as D'Alton's character might seem on the page, he was far from otherworldly, and a Google search depicted a man who fully inhabits this world in a kaleidoscopic, overachieving way, having more qualifications and commitments than he let on (plus a late Irish Cardinal namesake). Like all tinkerers I met, he was a committed polymath, and his tendency to meta-narrativise tinkering in authorial ways reminded me of another proselytising tinkerer, Mark Thomson (Chapter 9). Like Thomson, a couple of times he referred to his tinkered gadgets as 'props'. Some were literally: used as catalysts in role-playing games, the premise plots of which he writes. He described these games as if they were ethnographies: they're a *method*, he said, to experience others' personal struggles and situations in ways we can't through non-immersive, passive methods such as reading novels and biographies, or watching plays and movies. (Thomson, who also tells stories through his tinkered objects as well, also described tinkering as a 'gentle method of inquiry'.) In our second interview, D'Alton described his steampunk tinkering as a *method* 'which allows people to *live* their historical fantasies, rather than act them'. This resonated with Kate O'Brien's method of 'living' her historical narratives by crafting them.

Also living in Greensborough, Cliff Overton was my first research participant. He hadn't met D'Alton, but he knew of him. It's worth mentioning that several years have passed since I had conversations with these steampunk tinkerers, because movements (and people) evolve. Yet, collectively, they explain the constellation of cultures that maintain tinkering as a utopian practice in contemporary Australia.

When I met Overton, he was 36, a fast-talking and erudite hardware hacker who earned a living as an industrial-designer-turned-firefighter. I'd found him from a Google search for Australian steampunks. Greeting me at his immaculate front porch, Overton wasn't wearing his fob-watch, nor his waistcoat. And the home he shared with his wife Tanya, a civic planner, was nothing like the gothic, hovel-like *wunderkammer* I'd anticipated after reading his blog. A stainless new weatherboard perched on a modest precipice of bush, it housed homewares displayed uniformly on sheeny surfaces. This was a scene from a *Provincial Home Living* catalogue: there was nothing that hinted at subcultural disruption. (Later, reading Ken Gelder's *Subcultures*, I learned that hacker culture was never necessarily oppositional, nor necessarily 'dropout culture'.)

Perhaps sensing my disillusion, Overton explained to me that until recently he didn't know he was part of a globally bourgeoning culture called steampunk. 'It wasn't until I was at a friend's house,' he recalled. 'He showed me Datamancer [a technofetish website]. I found that there was a genre in which I could contextualise what I'd been doing.' All his life, Overton had been making utopian fictions – role-playing by tinkering. As a boy in the 1970s, he'd tinkered with timber off-cuts while his father built stage sets in amateur theatre productions. Into adulthood he continued to be 'fascinated by old used machinery parts', and started squirreling them away, with a dim ambition to 'one day see them reassembled into something'. One afternoon, he assembled parts into a 'robot cat'. From there Overton continued to mix theatre and invention, with a dash of inspiration from science fiction. Playing the role of the heroic lone inventor, he stepped into character as 'Mad Uncle Cliff', a neo-Victorian, top-hatted dilettante. Mad Uncle Cliff reimagined contemporary technologies – computers, desk lamps, radios, mobile phones – as if they'd been made in the 19th century. He refashioned these gadgets into elaborate contraptions. With shameless disregard for the form-follows-function efficiency that once guided his alter-ego's industrial designs, Mad Uncle Cliff heeded no laws to civilise the ornate tangle of cloth cables and copper wires, dials, brass pipes, cogs, pistons, mahogany and leather casing or blown-glass triodes.

By meme or coincidence, hundreds of like-minds across the globe were doing the same thing. Overton found hundreds, then thousands of unsung steampunk projects that were unleashed from lone inventors' back sheds and projected online to an international gaze. When he discovered this, he was 'intrigued … and also quite relieved that I was no longer on my own'.

When I started following this online party of tinkerers, it became apparent that many have almost identical stories to Overton's. In Massachusetts, for example, lived the pseudonymous Jake von Slatt, a prolific tinkerer who became one of the culture's biggest luminaries. *The Boston Phoenix* reported that von Slatt had accidentally unearthed Steampunk's online community. 'He came into it not realizing there was an entire subculture to define both his DIY habits and the visual aesthetics he'd long appreciated.' In turn, on von Slatt's *Steampunk Workshop* blog, a typical new visitor message was: 'I had no idea there was a name for what I am.' Months after I met Overton, Kate O'Brien (Chapter 6) told me that when she came across the movement, 'I felt like there was finally a word to describe what it is I do. Almost a justification.' At this time, 28-year-old Melburnian Bree Frost also told me: 'I didn't realise it, but steampunk was what I was doing.' And when John D'Alton stumbled upon steampunk: 'I'd been vaguely aware of the term "steampunk", but when a couple of years ago I read about it I thought, "Aha! That's the new term for what I'm doing".'

Collectively, these stories suggested that steampunk tinkering arose not from a commitment to shared political ideals, but from a convergence of personal impulses writ large in the context of globalised media cultures. As described in Chapter 6, steampunk is a utopian tinkering movement with fictional roots: a melting-pot of goth, science fiction and open-source IT movements. Like much tinkering practice, it's a double-act of dialectic extremes: a collective of individualists; an apolitical political movement; an ahistoric re-enactment; an anti-consumerist industry; a fictive reply to truths; and a whimsical response to serious issues. It's a belief that tinkering in your home can somehow change the world. For steampunk blogger Steam Dream, it's 'a thrilling revolt against the evils of mass-production'. 'For me,' writes the bestselling author Cory Doctorow, 'the biggest appeal of steampunk is that it exalts the machine and disparages the mechanization of human activity'.

Before the global boom of steampunk, its *aesthetics of decontextualisation* (discussed in Chapter 7) were already evident in ordinary suburban homes. Reproduction Victoriana can be found in the most conventional loungerooms and workplaces: in the commonplace homewares of Early Settler, Colonial Living, Copperart or any number of retail outlets, or in mass-media forms such as *Dr Who*, *Mad Max*, *Wild Wild West*, *Wallace and Gromit, The League of Extraordinary Gentlemen* or *The Golden Compass*, and in Japanese anime, *Steamboy* and *Howl's Moving Castle*. Steampunk's version of Victoriana is less reproduction and more revisionist, often characterised as nostalgia for a history that never was – but could have been: 'We are rebuilding yesterday to ensure our tomorrow.'

A month after I met Overton, I travelled to Melbourne's CBD, considered Australia's steampunk capital, to meet members of the Antipodean League of Temporal Voyagers. Contacting this group from its website, I was answered by Omega Howell, another steampunk tinkerer, who agreed to arrange a meeting with several steampunk characters. I first saw Howell crossing Melbourne's King Street with her then-12-year-old daughter, Melody. Both stood out boldly from other pedestrians. Howell wore an elegant, full-length black Victorian couture dress with bustle and red trim, and her hair was gathered behind her neck in a smart fishnet hat. Melody wore head-to-toe red: hat and dress coat. Among her many life-roles, Howell was a re-enactment specialist and technology educator. She now made Victorian clothes and costumes.

Howell had put the word out on the League's site a couple of days earlier, arranging for us to gather at Lockworks, on the first floor of a Hardware Lane building. I'd imagined Lockworks was a bar or lounge, or maybe a café or a nightclub. It turned out to be a hairdressing salon. Up its dim narrow staircase was an advertisement for the steampunk edition of *FIEND*, with the oxymoronic tagline 'Australia's Ultimate Dark Alternative Magazine'. This was at odds with steampunk rhetoric, which routinely rejects the 'dark' label, though many grew into the movement from goth. 'We're goths in brown,' Howell would later quip to me. 'Steampunk is not dark and spooky,' insisted Robert Brown, the lead singer for Abney Park, in *The New York Times*. Former goth Dan Turner-Chapman told me: 'Steampunk hits a more optimistic note, embracing the role of artisan.' A steampunk blog entry explains that goths are creepy because of vampiric sympathies, 'Whereas steampunks are – what? Weirdoes who take pocket-watches too seriously?'

At the counter, Howell introduced me to 37-year-old Lockworks proprietor Cass Edwards, all cleavage and cinched waist and dreadlocks, like some kind of punky Dickensian barmaid. She greeted me heartily. Soon history graduate Dan Turner-Chapman arrived in an ensemble of checked trousers, grey herringbone waistcoat and frock coat, and a William Morris-design tie. At 23, he was eloquent and recently married. 'She's not steampunk,' he said of his wife, 'but she does good Victorian'.

It was then I first met D'Alton, who arrived in a dapper Sherlock-Holmes outfit. He'd brought along an ornate musical shadow-box he was tinkering with, for his god-daughter. After him was 33-year-old Josh Orth, in a red-blazered military outfit with metal epaulettes and rank badges. These, and the fobs in his pocket, were made from salvaged computer parts, 'an occupational perk for me' – as Orth was the IT manager at the pathology department of the Peter MacCallum Cancer Centre. He listed *Flashman* novels, Jules Verne's *20,000 Leagues Under the Sea* and the film adaption of *The League of Extraordinary Gentlemen* as inspiration. Of his clothes he said: 'I'm going to have some embroidery done with unit markings and the like, all fictional and home-grown. I've been thinking of names for it, and I may have settled on The Third Imperial Viators.' Soon, 21-year-old web designer and law student Connie Chen showed up, wearing an underbust corset over her swashbuckling shirt. She produced a home-made plastic raygun that generated sparks, and everyone cooed over it as relatives might fuss over a newborn. 'Awesome,' someone muttered.

Clearly, these people tinkered with their self-styling as much as with objects. We pulled up chairs and sat in a circle discussing steampunk in what rapidly became less an interview and more a *salon*: in the Parisian – not the hairdressing – sense. The discussion turned to 'authentic' steampunk and questions of selling out. 'What attracted me to the steampunk community is the sharing of DIY knowledge,' said Orth, and there was an *Absolutely!* all around. Typically, steampunk tinkerers share methods through social media. The bestselling author Bruce Sterling told Enschede's 2008 Gogbot festival: 'If you meet a steampunk craftsman and he or she doesn't want to tell you how he or she creates her stuff, that's a poseur who should be avoided.'

'I suppose my take on steampunk,' said Turner-Chapman, 'is cultural, as a combination of historical and "punk", that is, socially-critical literature, and finding practical applications of those same principles'. Authentic steampunk, to him, had to be 'in some form an exploration of technology and its past and potential uses. In keeping with the optimistic side of steampunk, it has to be sustainable and efficient. Steampower for instance with the right fuel is much more efficient than internal combustion.'

What, I asked, about non-functional objects: ray-guns, goggles, fashion get-ups? Isn't steampunk just whimsy? A resounding *Yes!* drowned out other mutterings. That steampunk is play and imagination – that it is literary, speculative and utopian – is what gives it a political dimension, said Lockworks proprietor Cass Edwards. 'If we go back to Victorian technology and the industrial revolution and think maybe we made some mistakes back then? Perhaps there were particular developments in civilisation that we might have screwed up, and in a way we're reimagining what might have happened if we'd gone forward without mass production and the facelessness of modern society.'

Soon 28-year-old Bree Frost joined us, eating lunch out of a takeaway container, looking equal parts hippie and cyberpunk. At some point 32-year-old technical writer Anna Vesperman also slipped into the salon, in a full-length violet velveteen skirt. Her pale face was obscured beneath sunglasses. Most at the *salon* were tinkerers – not all of gadgetry, some of garments. 'I'd challenge anyone to draft a bustle gown and tell me it's not engineering,' said Howell. The essential concerns, she said, are the same. Many open-source crafters assert that digital technologies and 'fabriculture' are inseparable at many levels – and source-code is analogised with knitting technique.

I offered *Design Observer*'s criticism of steampunk tinkering as bourgeois, politically quietist, suburban and amateur. A round of chuckles erupted – no-one seemed offended. 'In steampunk,' said Turner-Chapman, 'there's no division between high and practical art'. Steampunk, this group told me, is proudly amateur and thus anti-elitist, a popular cultural form that doesn't demarcate borders between folk art, commerce and political activism. One 2013 steampunk manifesto framed this ethos as follows: 'The entirety of knowledge could be almost apprehended by a single individual [in the Victorian era]. There were still frontiers. There were fewer laws and governing bodies. Who wouldn't want all of those things back?'

Most of those gathered at the salon had at one time or another been involved in goth fantasy or historical re-enactment. Most worked in IT (some part-time); most lived in the outer suburbs; most had postgraduate qualifications and divergent career-changes. All were science fiction or fantasy readers and clearly all idealists. Only some had met, but there was an easy camaraderie, as they were familiar from online forums. With the exception of Vesperman, who was articulate but reserved, each possessed the fast-talking eloquence I'd seen in Overton, and later D'Alton. They shared these tinkerers' views that steampunk produces community through production, and in doing so can reorganise economies, and lead to social reform.

According to this *salon*, steampunk philosophies were now evident everywhere: in the steamy ideas, steamy aesthetics or very steamy styles that are now popular: William Morris wallpaper, cloaks from Sportsgirl, distressed typography in graphic design, Liberty fabrics, the Tesla car, the Linux movement, open-source generally. What's *steamy* about online technologies, said Turner-Chapman, is that artisans can now use them to reach a global market from their loungerooms. Howell added: 'Cottage industries can now be sustainable, as they can sell one-off items to a global market. By creating demands we are making, literally *making* the future.'

Free trade, Turner-Chapman added, 'is different from capitalism', because it has allowed the development of fair trade, an idea which 'is very steamy'. Steampunk gadgets are frequently commissioned, he said, and there is bourgeoning global steampunk trade on eBay and Etsy. If the market embraces steampunk – if well-heeled consumers pay premium prices for a fictional contraption's uniqueness and 'authenticity' – then all the better for the movement's prosperity. Such prosperity, as this group told it, is the antithesis of rapid-output sweatshop values that grease the wheels of exploitative capitalism.

This was much the same revolutionary promise generated from Silicon Valley entrepreneurs for more than a decade. Adrian Bowyer, who launched the first open-source 3D printer project (or 'citizen robotic') RepRap in 2004, invoked Karl Marx in his promise that RepRap:

> will allow the revolutionary ownership, by the proletariat, of the means of production. But it will do so without all that messy and dangerous revolution stuff, and even without all that messy and dangerous industrial stuff.

So people in their homes were promised the triumph and romance of revolution without the mutiny and bloodshed. Most utopian maker movement rhetoric coming out of the US focuses on individuals using additive technologies like 3D printers, and other digital fabrication technologies like laser-cutters.

These technologies offer exciting possibilities for the home tinkerer – especially the bots you can install at home or use in your local community Hackerspace, allowing you to download and print the open-source component that'll get your broken espresso machine working again, saving it from landfill, giving you a sense of agency, and changing the nature of product distribution and consumption. Some of the 3D printing polymers, though, are yet to become sustainable, and an online peruse of sites such as Cubehero, Thingiverse and Pinshape suggests many people instead tend to download useless plastic novelty trinkets destined to end up in landfill. Finnish researcher Cindy Kohtala is among those who've found that many FabLabs and Hackerspaces that provide community use of these bots pay little practical regard for environmental and social concerns. And most of these new technologies, platforms, systems, patents and spin-offs remain controlled by traditional multinationals, including Xerox, Hewlett Packard, Stratasys, Microsoft, and General Electric.

Still, these technologies have epic potential to those of us with digital access, and, as Australian journalist Guy Rundle has noted, their uptake has the potential to abet real social change. In *A*

Revolution in the Making: 3D Printing, Robots and the Future (2014), Rundle promises:

> the material revolution, a transformation in the production and distribution, of, well, everything. There will be enormous implications not just for Australia, but for the global economy, international relations and the fundamental structures of our lives.

Rundle is no utopian fantasist; he believes this revolution 'offers no clean break, no way in which we could simply shift into another mode' and involves 'messy systems with multiple layers of action'. (In fact, Rundle objects to what he calls 'the fetishisation of making'.) Years before Rundle, scholars were making comparable claims that are yet to eventuate. Sociologists Thomas Birtchnell and John Urry, for example, predicted that tech-enabled developments might 'offer possible futures of rapidly demobilizing global manufacturing, distribution and production' and 'of personal manufacturing disrupting existing global supply systems'. In *Desktop Manufacturing: A Home Brew Industrial Revolution* (2009), social theorist Kevin Carson wrote:

> Neighborhood workshops, desktop manufacturing, household micro-enterprises and online horizontal networks of peer producers will come to dominate the informal economy that must arise as the state-allied formal economy stagnates and decays.

'It's impossible', Carson asserts, 'to underestimate the revolutionary significance of this development'.

Most revolutionary narratives surrounding these technologies tend to turn social problems into techno-fixes and heroic quests. Recent (mostly American) maker movement books almost invariably invoke the values of Marx, but they also conflate democratic and egalitarian ideals with patriotism and commerce. In *The Maker Movement Manifesto: Rules for Innovation in the New World of Crafters, Hackers, and Tinkerers* (2014), TechShop CEO Mark Hatch writes: 'We have only just begun to see an outline of [the maker movement's] eventual power to remake the United States and the world.' And in the Reagan- and Trumpian-titled *The Tinkerers: The Amateurs, DIYers and Inventors Who Make America Great* (2013), former *Rolling Stone* editor Alec Foege equates tinkering with entrepreneurial freedoms throughout the history of US production, describing a maker movement engaged in 'a disruptive act in which the tinkerer pivots from history and begins a new journey that results in innovation' that leads to entrepreneurial success. Foege believes tinkering is 'inseparable' from entrepreneurship and wealth creation (a claim not reflected in the everyday lives of Australian tinkerers). And in *The Wall Street Journal*, former derivatives trader Nassim Nicholas Taleb wrote: 'Tinkering by trial and error has traditionally played a larger role than directed science in Western invention and innovation … which is closely tied to entrepreneurship.' Technocratic ideals of 'innovation' and 'entrepreneurship' are antithetical to the lived experiences and tinkering values described to me during my research.

And these claims tend to be loaded with oddly evangelical momentum. *Make*'s Dale Doherty wrote in 2014 that: 'We share a common mission. We want people to see themselves as makers and producers' – a missionary approach in an echo-chamber of tech proponents. Mark Hatch, for example, wrote: 'Every revolution needs an army … My objective with this book is to radicalize you and get you to become a soldier in this army.' All heavily vested in 3D technologies, Xerox, Ford, Hewlett Packard, Stratasys, Z Corporation, Deloitte, and General Electric are spinning the same democratising and revolutionary claims into grand narratives of production history. Ford's Open Innovation Manager told a 2013 symposium that:

> tools for innovation have been so rapidly democratized once again … coming together now in incredibly democratized ways … Like Jeff Nelson, who is one of my colleagues, actually created [with a 3D printer] a shift knob for a car, incredibly cool … This is going with a way back to Henry Ford and our learning. Henry Ford spoke about making the desirable affordable. We can now make that desirable, in your mind, affordable. And it can be democratized …

Unlike steampunk tinkerers, most of these sources mount a narrative that historian W Patrick McCray calls 'the cult of the Great White innovator'. Once we 'redefine our sense of what an innovator is and what talents she might possess, we start to see that the industrial revolutions of the past few centuries did not have one single global meaning'. He writes that all the:

> economic reshuffling, social upheaval and environmental exploitation of modern industrial revolutions look very different from the perspective of a person living in Europe than from the perspective of people in Asia or Africa, for example. If we leave the shadow of the cult of the Great White Innovator theory of historical change, we can see farther, and deeper.

To McCray, a focus on the real and the everyday material practices can allow us to understand that 'innovation and technological change are more than *just* making things … this allows us to begin to glimpse a more familiar world where activities such as maintenance, repair, use and reuse, recycling, obsolescence and disappearance dominate.' A more global picture, he believes, would include the Lizzie Otts of the world, who invented things we now take for granted. Recall in Chapter 2, Lizzie Ott adapted a rear car-wheel to invent a mechanised washing machine. Rather than using her car as part of the travel revolution promoted at the time, Ott reinvented the machine itself for her own domestic purposes.

Deloitte, like other corporates, seems to be wary of everyday makers even as it celebrates them as potential markets. It charts the maker movement's success by tallying the number of Hackerspaces, techshops and online trading sites globally. Predicting that the maker movement's 'impact will be felt across more areas of society and business', it forecasts that 'it is only a matter of time before large firms begin to feel the impact as multitudes of niche products collectively take market share away from generic incumbent products'. So it advises business that makers are:

> a broadly distributed innovation ecosystem. They tinker, build, market, and sell unique items. The benefits of paying attention to this ecosystem and its output include the ability to observe emerging trends and technologies within the ecosystem. … Monitoring emerging trends can allow scale operators and orchestrators to better shift resources, not only to support but *also to shape these trends*. [my emphasis]

So clearly, multinational corporates aren't simply reporting passively on the maker movement: they themselves are shaping and promoting the narrative in a top-down discursive feedback loop. Tech proponents see themselves as *shaping* the movement, frequently championing 'disruption' and 'innovation'. (Recall Foege's description of tinkerers as people engaged in 'a disruptive act in which the tinkerer pivots from history and begins a new journey that results in innovation'.)

As technology critic Evgeny Morozov has observed, big-tech innovation and disruption narratives are self-serving and can be harmful, because they routinely inform public policy and funding. Morozov believes many of the tech giants spruiking these new technologies 'have already taken on the de facto responsibilities of the state' when it comes to regulating and distributing them. (This is manifest in corporate–government partnership funding of Australian fix-it lounges and hackerspaces, documented in Chapter 2.) This doesn't really serve the everyday interests of the home-based DIYer. According to tech journalist Leigh Alexander, the very word 'disruption' (like that of 'hacker') has become a startup culture buzzword that 'has the aftertaste of a sucked battery. It doesn't even mean anything anymore'. It simply seduces a closed-shop of technocrats.

In a 2011 *Utopian Studies* essay, ethnographer Nicole Dawkins found that a 'discourse of DIY' in maker culture tends to be framed by a neoliberal rationality in its celebration of consumer choice, entrepreneurialism, individualism and self-improvement. Likewise, within maker movement discourse, current 3D consumer technologies are routinely equated with these entrepreneurial freedoms and choices – using the same rhetorical tools as New Economy claims generated from Silicon Valley in the 1990s. *Disruption* is a term that's appropriate for everyday tinkering (think, for example, how Adis Hondo changed Australia's historical record with his spycam tinkering). But the prevailing disruption narrative is heavily male-dominated and so are the fabrication practices it describes. There's some inclusion of women commentators (such as Intel's Genevieve Bell and academic Anne Balsamo) but, in the main, the commentary effectively ignores half the world's population and their everyday practices and habits.

This is a disease in broader technology development throughout industrialised history, in which systemic exclusion of females is well-documented by scholars. (And even by popular news outlets. In a 2015 story explaining why some brilliant inventors fail to attract Silicon Valley venture capital for worthy and viable products, *Newsweek* declared: 'They don't have penises.') So it's also no surprise that within maker movement literature, tinkering tends to be equated with forms that are conventionally assigned to male gender-roles – robotics, tech inventions, mechanics and engineered structures. Even though Deloitte forecasting includes craft commerce sites such as Etsy, more traditionally female forms like fine handcrafts, clothing and food production aren't frequently characterised as 'tinkering' – their impact tends instead to be confined to 'slow food', 'craft' or 'artisanal' discussion.

But the tinkerers I studied are first and foremost practical makers who regard new 3D technologies as simply one fabricating option among many, and thus purely instrumental and optional for their own discrete projects (rather than for broader reformist ones). Few tinkerers I've come to know regard themselves as part of a maker movement; few have actively contributed to open-source fabrication projects: their knowledges were often too tactile, too specific or too tacit, and so the participatory cultures within which they engaged were very often confined to their immediate actual (not virtual) communities. Not one had the entrepreneurial zeal or ideals portrayed in maker books; none expressed interest in manufacturing (or producing) a complete product by downloading a design and printing it on a 3D bot or laser-cutter: this would defeat the very values of material tinkering. Most expressed interest in a bot's potential for prototyping or printing unfindable components: none had in fact used one. None expressed interest in modifying others' off-the-shelf designs on sites such as Thingiverse or Pinshape. Instead, their stories *refute* the nationalistic and entrepreneurial evangelism generated from dominant maker-movement discourse.

A month after the *salon*, I settled with a mug of tea into soft salmon leather-look armchairs beside a copper woodfire flue and flat-screen television at John D'Alton's home. He wore his funereal frock-coat that resembled his real-life cassock. 'Actually,' he told me, 'it's a tinkered Indian Nehru coat'. On our first meeting at the salon, he had assumed the character of Professor Uncle Festa (actual spelling) – not just a reference to Addams Family goth, but to his avuncular role in Australian steampunk. 'I'm a tribal elder,' he said, adding: 'in the sociological sense'.

More than figurative and speculative, D'Alton's tinkering had materialised in various cultures and genres. In India, where he, his wife Lyn (a former doctor, then a painter) and their children spent several years, D'Alton worked with villagers to build solar power 'out of old fluoro tubes and tin and black paint etcetera. It was fun.' Learning how to build solar panels from resources-at-hand, he said, 'inspired me to live more sustainably' back in the First World. At the time of our first interview, D'Alton was turning bits of broken bikes into a six-wheeled cycling contraption. He told me he'd been doing such things 'since I was a kid making things and collecting Victoriana and steam engines. I made a mechanical computer when I was about 15, and then at 16 built a kit computer when there were only about ten of us in the whole country with home computers, in 1976. I've been building wooden and metal and old-style things ever since.'

D'Alton had tinkered with music boxes, clocks, costumes and greywater filters. For the latter, he said, you can find the whole system by scavenging and allowing microbial forms to perform their miracles: all you need is containers, a washing-machine hose, and gravel: 'The bugs that form do all the work. They convert nitrites to nitrates.' He described tinkering as 'an achievable response' to the excessive, cheap flow of factory-made products that wind up in landfill and in our oceans: in an email, he told me he sees steampunk's green and inclusive values as consistent with orthodox Christianity. Other, non-steampunk figures have offered similar idealistic projections. To

Australian academic and industrial designer Miles Park, 'by tinkering, there is a definite connection between the life-span of products and tinkering. If you can slow down that throughput of stuff going into landfill, surely it's got to contribute to reducing that overall burden of landfill waste.'

D'Alton told me steampunk was a movement in which sustainable solutions are being generated through a Victorian sense of optimism and hope, a time when anything seemed possible: 'We haven't had that culture since the late 60s and 70s, when, after Woodstock and Nixon, hope for the future was lost. I remember when Gough got in, I remember when conscription was stopped ... all the hope of that era went away somehow.' I asked: *With the Dismissal, perhaps?* 'Yeah, with the Coup,' he corrected me. 'That and the counter-culture falling apart. Really, there hasn't been any broader social conscience to replace it.' What about recent anti-consumerist movements and the return of large demonstrations? 'It's all angry,' he said, adding: 'There's a valid anger, but it becomes all-absorbing and unproductive, and a lot of the counter-culture has gone to angry victimhood and powerlessness, and unhealthy identity exclusivism. Whereas steampunk is an optimistic and inclusive counter-culture.'

'The DIY rebellion,' he continued, 'is the happy radical as opposed to the angry radical, with an optimistic vision of the future. I don't think we've had that in a long time.' Might steampunk tinkerers have something in common with that other Victorian-era quiet reformist movement, Fabianism?

> Yes, personally I share much with the Fabians, and I suspect many steampunks would unknowingly. Many are not really aware of the political implications of reinvigorating old technologies and strengthening cottage industries. So yes, for some it's a quiet activism, and for others an unknowing political impact. I think some steampunks are more into creating a newer

> alternative society as much as reforming the existing one, but it's peaceful rather than [violent] revolutionary, the happy radical as opposed to Trotskyists.

Earlier, Cliff Overton had told me steampunk tinkerers had widely divergent political worldviews. D'Alton agreed: 'Some tinkerers are very right-wing libertarian.' My method of rapport most likely screened out tinkerers from the Right from the sample in this book, but there's plenty of evidence to suggest that tinkering spans the ideological spectrum. The Homesteading movement in the US, for example, is deeply conservative, and also deeply committed to many of the values of tinkering. One of the most important and influential thinkers for my research, philosopher and motorcycle mechanic Matthew Crawford, comes from a conservative background, having worked as Executive Director of the pro-Republican Washington think-tank, the George Marshall Institute. (He later disavowed the Institute's claims. His work there, he said, 'demanded that I project an image of rationality but not indulge too much in actual reasoning'.) And as this chapter has described, some of the fiercest advocates of tinkering are Silicon Valley entrepreneurs who, despite their progressive posturing, conflate market values with democratic ones.

There are also tinkering conservatives in the folksy, open-range style of the agrarian philosopher Wendell Berry, who could equally be interpreted as progressive. A libertarian with a skeptic's eye to the claims of big capital, Berry seemed to be what conservative US journalist Rod Dreher describes as a 'crunchy conservative'; someone who eats organic food, cares for the environment and homeschools their kids; and who also 'believe[s] the economy must be made to serve humanity's best interests, [that] big business deserves as much skepticism as big government … [who] lives by … moral norms necessary for a civilized life, and which are taught by all the world's great wisdom traditions'.

D'Alton told me that some steampunk tinkerers came to the movement with anarchist views, probably from cyberpunk, but that this is a movement where collectivists like him 'have much in common' with libertarians. The defining characteristic of steampunk tinkering, he told me, runs deeper than political tribalism. 'The one defining characteristic about steampunk is *doing* stuff,' said D'Alton. 'Making and *up*cycling. More than recycling: reusing things from the past and making them better than before. It's saving the environment from all that junk, and making it more beautiful than before.' What about Bunnings purchases, or plastic ray-guns? Isn't that selling out? 'It's also a pragmatic movement,' he said. 'There are always trade-offs between beauty, functionality, time and resources. As long as the broad worldview stays there, it's steampunk.'

Once I understood D'Alton as an ecumenical character, his fantastical methods, his pluralist persuasions, this dizzying array of time-travel and wizard-chaplain-tribal-elder role-playing – all of them made complete sense. They're all simply different versions of the same story. *Star Trek*, *Dr Who*, Ghandi, Islam, John Ruskin, the wizard, Christianity, *Babylon 5*, steampunk: they're all narratives of utopian revolution, epic struggles in which the quest might seem impossible, but in which good prevails where hope and faith are kept alive. Block's story (Chapter 3) showed the ways tinkering is the embodiment of hope. Hope and faith were D'Alton's professional stocks-in-trade, and he exported these into his tinkering domain. To an email in which I posited craft movements' reformist failures throughout history, he replied:

> I am inspired greatly by Ghandi who never gave up hope and eventually succeeded beyond anyone's wildest dreams. Yes, some aspects of his dream never happened, but so much did that would not have if he succumbed to defeatism … People will find new ways to change the world even if previous ways failed. The creative frustration evoked by failure should not be rationalised but transformed into new action.

To D'Alton, as to the Arnhem Land people in Adis Hondo's chapter, the material and the spiritual are inseparable. Once I immersed myself in his thinking, tinkering, too, made sense as a revolutionary impulse of hope and faith. The everyday strategies of tinkering are not in themselves always overtly idealistic or spiritual but are grounded in particular, concrete, personal, realpolitik and compromising solutions to material, environmental and political problems. Utopia, likewise, isn't about transcending personal, practical, here-and-now outcomes: it's understood (and dismissed) as mythmaking, but it's always attempted in material ways. To cultural scholar Jonathan Sterne, *techne,* too, is at once creation and contingency, holding an 'ambiguity between the actual and the possible'. Techne is born of a continuum of social habits and practices: socially and individually embodied knowledge that constantly moves between possibility and actuality.

Michael Drinkwater's story, briefly mentioned in other chapters, offers a salient example. In 2000, Drinkwater's then-Gippsland-based family relied on wind and solar technologies to power their home. They installed a backup generator that ran on expensive and polluting mineral diesel. Drinkwater was inspired by the story of an American couple, Josh and Kaia Tickell. The Tickells were filmmakers and sustainable technology entrepreneurs who famously made their way across the expanse of the US over two years, running their van on biodiesel using a home-made processor and the waste of fast food outlets. Drinkwater read the Tickells's instruction book, *From the Fryer to the Fuel Tank*, available free online. The making of biodiesel involves readily-available cheap materials: any vegetable oil (including used oil), methanol, and caustic soda. Small amounts of biodiesel can be made with basic kitchen appliances, but Drinkwater wanted a machine that could manage quantities that would allow him to meet all his fuel needs, including his tractor and truck that towed a horse-float. So to make his own biodiesel processor, he found an old dishwasher and converted it into a reactor. The dishwasher was 'almost perfectly suited to this purpose as it had a high wattage heating element' of 2,000 watts. It also had a 'powerful recirculating pump for mixing the reactants' and 'a capacity about the same as my truck and generator fuel tanks combined'. Drinkwater removed the machine's dish racks, spray arms, door springs, and insulation from the back of the cabinet. He fitted a gas burner ring to preheat the oil and 'keep the concoction warm after reaction while settling, as is necessary during cold weather'. The biodiesel was then pumped to an overhead storage tank, and the glycerine by-product was drained off. Drinkwater's invention 'met my entire fuel demands with biodiesel'. Since he bought the methanol and caustic soda, it worked out to about 13 cents per litre (as I write, mineral diesel is about $1.60 a litre).

Like every tinkerer in this book, Drinkwater saw this tinkering as revolutionary: not just as thrifty good sense but as an antidote to the personal and political discontent wrought by the conditions of mass production and modernity. He saw it as a way to make concrete change in areas where policy makers are incompetent. He later wrote in an email: 'Tinkering is taking back control of my environment … We are being channelled into reliance on external intelligence, bypassing our own, which leads to exploitation, control and loss of skills.'

This was precisely John D'Alton's argument, the idealistic arguments I heard at the salon, the arguments in writings about Luddites and about John Ruskin and William Morris, and also the argument fictionalised in Cory Doctorow's novel, *Makers*, whose revolutionary *mise-en-scene* is now generated in Silicon Valley maker-movement discourse. It was the argument of futurist Alex Pang, and of historian Kathleen Franz, who wrote that everyday suburban car tinkerers last century were 'challenging dominant ideas about who could access and have some power over new technologies and who could cross the boundaries between consumption and invention'. In an issue of *Utopian Studies*, blanket-maker Joseph Travis argued along the same lines. He wrote: 'If we take control of production, we are taking that power back. This is a path for those of us who would nonviolently change society: change its commodities, its understanding of production, of distribution and exchange. Even the way we relate to things personally.'

I wondered whether, as a magic man, 'tribal elder' and religious chaplain, D'Alton would identify with the secular descriptions of magic I describe in Chapter 3. I emailed him a draft. In this chapter, Chris Block's character is, literally and figuratively, a far cry from the fabulist story of steampunk, but John related to Block's story so much that he emailed the draft back to me, riddled with highlighted areas for us to discuss. 'I resonated with so much of it!' he wrote, 'Especially the magic moments and Gell stuff.' To me, Gell's theory of magic is atheist and ruthlessly reductive: certainly not spiritual. But on my next visit, after loosening his chaplain's collar and cassock (he'd dashed home from a meeting to lunch with me), D'Alton said: 'As I read it, I kept thinking "*Exactly*".' He told me Block's magic tinkering experiences, simpatico with his own, could be interpreted as 'a manifestation of God'. Feeling awkwardly secular, I didn't question him about this, but our further email exchanges suggested the comment was as much about the magic of tinkering's storymaking as it was the transcendental grace of material engagement described in Chapter 4. Material tinkering seemed like a method of transubstantiation – the material writ spiritual. Driving home from my interview with D'Alton, I heard a radio broadcast in which theologian Neal Plantinga said: 'Jesus built the kingdom as a carpenter before he built it as a rabbi.' (Once home, I Googled this quote – it can be found in many sources, including Mike Erre's 2008 book, *Why Guys Need God: The Spiritual Side of Money, Sex, and Relationships*.)

It seemed fitting that D'Alton, a chaplain and theologian, would see material production in a peacefully revolutionary political schema. According to Sennett, Christianity from its origins has placed great emphasis on Christ being the son of a carpenter, and craftsman saints appeared in the Middle Ages because Christianity embraced materially productive work, as 'these labours could counteract the human propensity for self-destruction … craftwork was peaceable rather than violent'.

Early in my research, I interpreted steampunk as a form of neo-Luddism. D'Alton, along with others at the steampunk salon, readily identified with Luddism, the world's first organised labour-movement, long misunderstood as an anti-technology movement. In *Against Technology*, literature professor Steven Jones asserts that the original Luddites were in fact themselves skilled and enterprising technologists – they used complicated looms, heavy hand sheers, and large cropping and weaving machines. In *Rebels against the Future: The Luddites and Their War on the Industrial Revolution*, Kirkpatrick Sale argues that the Luddites of Regency Britain were never opposed to technology, but to 'Machinery hurtful to

Commonality' that robbed workers of their senses of agency. Nor, he argues, was Luddism a conflict between old and new – it was one between custom and commerce. Weavers, combers and dressers of wool and cotton did not set out to be machine-breakers – they simply objected to being reduced from 'self-respecting artisans, with long traditions of autonomy and status, to dependent wage slaves'. Similarly, with its 'Love the machine, hate the factory' motto, the deeply nostalgic *Steampunk* magazine champions a humanist movement that 'seeks to take the levers of technology from those technocrats who drain it of both its artistic and real qualities'.

To literary studies scholar Elizabeth Outka, new hybrid forms, 'both commercial and literary', are arising out of a merging of novelty and nostalgia, of commerce and a sense of authenticity that is 'allegedly free of the vulgar taint of commerce'. The 'commodified authentic,' she argues, 'provides a model for understanding some of the contradictions *of* modernism as inherent *to* modernism'. To Outka, the paradox of the commodified authentic is its very appeal, allowing people to connect to values they see as authentic, and yet be fully modern.

To anthropologist Susan Stewart, nostalgia is simply a 'narrative utopia' based on longing and desire which 'has only ideological reality'. To cultural theorist Svetlana Boym, nostalgia is an expression of utopianism; it turns history into a private or collective mythology in the face of the alienations or atomisation wrought by globalisation and modern ideas of time, progress and nationhood. It can project a 'phantom homeland' and 'modern reinvention of tradition' which is 'retrospective but also prospective. Fantasies of the past determined by needs of the present have a direct impact on realities of the future

... nostalgia is about the relationship between individual biography and the biography of groups or nations, between personal and collective memories.'

Fictional narratives of history often remain more vivid in people's minds than more scholarly versions of history (see, for example, Kusima Korhonen's 2006 book *Tropes for the Past: Hayden White and the History/Literature Debate*). Utopianism, a kind of 'speculative myth' that envisages social change, is inscribed into steampunk and other artisanal products, but no less into mass-produced ones. According to design philosopher Jonathan Chapman, fiction is embodied in everyday consumer products. In *Emotionally Durable Design* (2005), Chapman argues that the fictional stories embedded in consumer technologies signal dominant utopian values of the time. Refrigerators' glossy facades fictionalise a cleanliness that belie the situation within; 'air-intakes' on the side-panels of a BMW ZI spin yarns about high-performance contemporary aerodynamics; the handles on the corners of 2000 G4 Macs created 'a fictitious caricature of nomadic urban mobility'. Most of these products give us 'a utopian, idealized and slightly fictional view of how the world could be; they are fictional in that they depict imagined futures – alternative versions of reality – that users feel compelled to pursue and engage with'.

Yet the narratives in these consumer products can be short-lived because these products are rapidly 'de-fictionalized ... once fictions are explored, demystified and known ... new fictions are sought'. For a product to have long-term emotional endurance, according to Chapman, it must 'possess richer, lengthier and more complex fictions if the consumption process is to be satisfying and longer lasting'.

So if there is depth and breadth – or a backstory – to the way products are experienced, imagined, theorised and fictionalised, this is one way to prevent redundancy or disposability. D'Alton told me that when you have meaningful relationships with objects, you're less likely to consume without mindful criticism, or to throw out things with which you have formed relationships 'and so you prolong the life-cycle of things'. He caressed an unusual bookcase in his dining room, pointing out how his friend used wooden pegs to construct the shelving. 'Every time I pass this bookcase,' he said, 'I appreciate it. It has personality, and a story attached. You establish community and relationships from things made by people.' In contrast to his Ikea shelves downstairs, he said, 'I'll never throw it out.'

Sociologist Pascal Gielen has criticised the academic, scientific and political taboo concerning everyday utopian thinking. He believes this 'destroys the liveliness of the political and the civil space, especially that of the *social imagination*'. The current dominance of so-called 'real-world' or 'pragmatic' politics 'deprives politics of chances for developing a long-term vision. Nowadays, any visionary project with an eye on an ideal society invariably runs aground upon the realpolitik of budgetary policies.' In *The Age of Consent: A Manifesto for a New World Order* (2003), George Monbiot tells us that the French Revolution, female enfranchisement, the fall of the Berlin Wall, the rise of communism, the fall of communism, and the aspiration of decolonisation movements all were described as 'unrealistic' just a few years before they happened. And they all relied on social fantasy and imagination. But politics, writes Gielen, 'has become policy, and governing a matter of bookkeeping. This corners the imagination, or rather: sends it into exile to the exclusive domain of fiction.'

Fiction requires we suspend our disbelief, while redemptive narratives expand our hope. Psychologists believe our social behaviour and our ethics can be charted by such narratives. In *The Redemptive Self* (2005), psychologist Dan McAdams set out to find out what makes 'generative people' with exceptional caring and commitment to the wellbeing of others. What mobilises people into caring behaviour? He found that such people tended to identify with or locate themselves within redemption narratives. These narratives could be religious ones, or else of their memories of healing.

Tinkerers routinely generate redemption narratives (discarded doors become iSlate; disused WW2 storage tank becomes home; abandoned dishwasher becomes biofuel processor; junk becomes *Titanic* museum display). The bestselling US author Michael Pollan has pointed out that the narratives that attend artisanal products implicitly criticise the values of the large-scale government-industrial complex, offering practical, material salvation: *home-made, sustainably-grown, single-origin, fair trade, locally-produced, shade-grown, grass-fed, dolphin-safe, hand-crafted, free-range, small-scale, family-owned, original-recipe.* According to Chapman, meaning-making engagements with products increase their 'emotional durability' and thus their sustainability – waste being a symptom of 'expired empathy'.

So if we make and engage deeply with material products, we not only inscribe our own utopian narratives into them – we can also recognise and criticise prescribed utopias inscribed in them by others. To Crawford, this involves critical thinking 'from the logic of things rather than the art of persuasion'. Accordingly, thinking 'materially about material goods, hence critically, gives one some independence from the manipulations of marketing'. If you understand coal's inherent proclivities and life-cycles, then 'clean coal' political campaigns won't wash with you; if you understand the values of craft, you won't be duped by corporate craft-washing campaigns.

This means the material knowledge gained by tinkering can be a way to question the fantasy projections embodied in commodity products and their attendant marketing. To futurist Alex Pang, this is one of tinkering's essential powers:

> Tinkering … encourages a kind of spirit of open-ended inquiry about the material world, and it also encourages a kind of empowerment, a kind of sense that you don't just have to take the world or take things as they come, but rather that even in today's apparently high-tech, slick, very pre-processed, very produced world, it's possible still to break open the covers, to get into the silicon, to get into the gears and to improve them, and to make them your own is a very powerful message.

The people I interviewed understood tinkering as an inherently critical activity that continually questions the basis of its own value, while also testing theory and practice. To D'Alton, this was why tinkering 'is a question of morality'. He said tinkering isn't simply the joy of tooling with objects – it's about how we choose to live, relate to materiality and to the planet. The planned obsolescence and imposed fictions of efficiency embedded in many new material products, he told me, amount to 'deliberate immorality. It takes away people's sense of self-sufficiency. It wastes resources, it wrecks the environment, and it creates dependency through certain kinds of consumer practices that are the norm in capitalist societies.' By fooling around with objects, tinkerers are making their own versions of utopia. By custom-building things, they're crafting their world, and custom-building meaningful lives.

9

Exhibition

When I pressed Mark Thomson to describe his vocation, he said he was a 'leftie trouble maker and cultural bomb thrower.' As such, he founded the International Festival of Explosions, whose poster promised that jet engines, bombs, muffler-free race-cars and ship foghorns would erupt in the tiny South Australian town of Terowie, 'all at the same time!' (Thomson had to explain this was a prank when people queried in earnest.) Later, he described himself as a 'proseletyser' of tinkering. It was hard to peg him. When he graduated from the South Australian School of Art in the 1970s, his career ambled into graphic design, agitation, political screenprinting, copywriting for the labour movement, ministerial speechwriting, music performance, and theatre prop-making. During my research, he quit his part-time speechwriting job for a South Australian ALP minister to concentrate on writing and design of the ABC television series, *The Lost Tools of Henry Hoke*. His other jobs included researching Australian 'networks of resourcefulness', producing museum commissions and working with Men's Sheds and Big Picture Education. With two friends he ran the blokey Institute of Backyard Studies (IBYS), 'home of shed culture'. He spoke at local inventors' events and at the World Maker Faire in San Francisco, after which he helped found the Adelaide Mini Maker Faire in 2013.

A couple of years after I met him, Thomson was given a grant from Intel to produce material projects with open licence: 'Here's me taking patronage from Intel,' he said. He wasn't entirely sure 'why they'd want to hire somebody like me' – he supposed this was 'to give them kind of insights into other ways of doing things … Am I

Mark Thomson's Twittering Machine (made with Phil Gerner) being trialled by festival-goers in Melbourne, 2014.

dealing with the Medicis of the twenty-first century? I probably am. This is the modern pope.'

He had an encyclopaedic knowledge of tinkering. Working in the labour movement exposed him to people who approached their lives with a kind of creativity 'not bound by what was officially accepted as art'. In the 1990s, Thomson started carrying his camera and notebook into people's sheds and workshops. Inspired by the work of social historian Studs Terkel, he published his high-end photography and reportage in the popular-market *Blokes and Sheds* (1995), 'a Father's Day book' that 'became a national bestseller and made my life hell for a while'. This was followed by *Stories from the Shed* (1996) and *Meat, Metal and Fire* (1999) which documented 'backyard barbecue engineering' – everyday improvisations from domestic or industrial objects including wheelbarrows, car-parts, lawn mowers and garden shovels.

Thomson's next books championed useful crafts as an unsung and classless creative force. His philosophy iterated a long lineage of craft-trade proponents (among them, Denis Diderot, John Ruskin, William Morris, John Dewey, Josef Albers, Walter Gropius, Peter Dormer; and in Australia, Robert Hughes and Peter Timms). The popular-market *Rare Trades* (2002) focused on a horologist, oculist, milliner, wig-maker, shipwright, shingle-splitter, bullocky, blade-smith, quarryman, stone-letter cutter and others, and became a National Museum of Australia touring exhibition. *Makers, Breakers and Fixers* (2007) profiled skilled Australians of a range of classes and outlooks. These weren't the kind of people usually considered 'craftsmen'; nor were they part of the knowledge-work cohort understood by theorists such as Richard Florida as 'the creative classes'.

During all this, Thomson developed a deep disdain for cultural gateposts defining fields of creativity. The high art scene 'left me uneasy. I found the plutocrats who constituted the art market a repulsive bunch'. On the other hand, he found the community arts movement 'worthy but pretty grim and humourless'. The arts, he believed, were cloaked in 'subtle barriers of privilege' in which 'the gatekeepers require a weird Masonic handshake to let people in'. (Later, he qualified: 'I meant that symbolically, not literally'.) Most of his art-school contemporaries, he said, 'are now major players in the funding world. It's emperor's new clothes stuff. All the participants can't see their irrelevance to the wider world.' In many conversations, Thomson said creative production should be recognised as part of everyday life. He said that other cultures such as India had no public galleries until the middle of last century, and so 'hadn't made art galleries into a sort of sump into which all creative activities could be drained and concentrated'.

But during my research, Thomson's own tinkering would be absorbed into these sumps. His career took a fateful direction when, during research for *Rare Trades*, he chanced upon a couple of artefacts manufactured by Hoke's Tool Company, or more specifically by the 1940s tinkerer Henry Hoke, whom he later described to the ABC's *Stateline* as 'probably Australia's greatest unknown inventor'. He told *Stateline* that Henry Hoke 'was on very good terms with Einstein: Einstein signed letters [to Hoke] "Bert"', and later told ABC radio that Hoke was 'Australia's very own Thomas Edison'. Rummaging through a garage sale, he said, he found 'giggle pins and chuckle valves in a couple of old boxes, and I'd never heard of these things'.

So Thomson delved into the history of Hoke. He began collecting Hoke's inventions. Many had no apparent use, but had names like 'leg pulls', the 'long weight' and 'the learning curve'. Thomson began restoring, collecting and documenting Hoke's Tool Company artefacts – spark plug sparks, threadless screws, pointless nails – and when his publisher HarperCollins learned of these, it commissioned a book to document them. After *Henry Hoke's Guide to the Misguided* (2007) was published, *The Lost Tools of Henry Hoke* became a touring exhibition in public art galleries and museums. As I began research, it was a year-long exhibit at the National Science and Technology Centre Museum in Canberra. As I finished my research, *The Lost Tools of Henry Hoke* television series (narrated by Bryan Dawe, Dick Smith, Lucky Oceans, Maggie Beer, Robyn Williams, Kosta Georgiadis, Kevin Sheedy, Amanda Vanstone and other prominent figures) was broadcast nightly before the 7pm news.

Despite his material legacy, Henry Hoke is a fictional tinkerer. Hoke's Tools is of course Thomson's riff on *hoax tools* ('There's no tool like an old tool', his site explains). That Hoke is an invention isn't always immediately apparent to exhibition attendees or television viewers. In one email, Thomson wrote that narrator Bryan Dawe 'kept coming across people quite convinced Henry was real, as did Dr Paul Willis, Director of RiAus and former presenter of ABC's

Quantum. And I got heaps of such people on the Facebook page.' Yet some of the institutions that exhibited and broadcast Hoke were being played as much as their audiences. When I asked Thomson whether, considering his contempt towards the fine arts bureaucracy, he felt conflicted about displaying Hoke's inventions in public art galleries, he replied 'Not hugely because I don't care.' He added: 'I can't stress enough that I couldn't care less about the art world. In some way I feel I'm sneaking a few things under the radar. They're seeing it as art because it's in an art gallery.'

Certainly, Hoke's engine-oil sensibilities would seem to have more currency in shed culture than high culture. In Chapter 6, when Kate O'Brien describes her incredulity at her everyday tinkered objects being treated with white-gloved reverence once in the gallery space, she seems to have a kindred sense of imposture. In creating and promoting Henry Hoke, Thomson had trodden on what intellectual property lawyer and novelist Stephen Gray calls 'a fine line between forgery and imagination' that has characterised much notable Australian cultural production since white settlement. As Simon Caterson explains in *Hoax Nation* (2010), hoaxing is 'inseparable from the wider Australian narrative'. Unlike James McAuley and Harold Stewart, who famously invented the Modernist poet Ern Malley with the express purpose of disrupting cultural conceits, Thomson didn't have explicit premeditative intent. But Hoke certainly sits in the cultural tradition of Malley, and later Walter Lehmann – like them, Hoke exploited cultural institutions to subvert their authority. (Lehmann isn't as famous as Malley. In 1961, the poet Gwen Harwood submitted two sonnets by 'Walter Lehmann' who was 'an apple orchardist in the Huon Valley in Tasmania' to *The Bulletin*. She later described the poems as 'poetical rubbish' that would 'show up the incompetence of anyone who would publish them.' Read acrostically, the sonnets said: 'So long Bulletin' and 'Fuck all editors'.) The figures of Malley, Lehmann and Hoke are all iconoclastic; all whistle to an anti-intellectual register; all are opportunistic and referential; all require their authors' keen attunement to certain cultural idioms to pull off their pranks.

And all occupy a narrative that's as much cultural as individually -authored. Thomson told me Hoke's lineage spans back to actual Australian trade pranks. According to his (non-fiction) *Rare Trades*, a longstanding trade tradition 'was the runaround or wind-up'. An apprentice might be instructed to go into the storeroom and search until he finds the long weight, or to fetch the left-handed screwdriver, sparks for the spark plugs, spots for the spot-welder, a missing skyhook, and so on. These sometimes humiliating pranks, Thomson explained, are traditional initiation rituals that might be today seen as workplace bullying. 'I first heard about the phenomena of apprentice tricks in the trade union movement years ago but didn't take much notice. When I was doing the *Rare Trades* book they came up several times and I made the first bogus packages. I think it was a box of spark plug sparks. People liked them so much I just kept making more and more of them.' Earlier chapters describe the ways tinkering has always been associated with fakery, fantasy and the crafty trickster. Scholarly searches, too, throw up article titles like 'Tinker-centric Pedagogy in Literature and Language Classrooms', in which tinkering is promoted as a method of creative writing and thought.

Before I learned about Hoke, I told Thomson I was unclear about what type of maker he was. He replied: 'I don't know what kind of maker I am either. It's something to do with telling stories and designing.' He said the only times he considered his identity was when I probed him with these kinds of questions. His tinkering occupied a realm in which categories were meaningless. When I first visited his home, a 1960s fibro house in a leafy street in the Adelaide Hills, among his inventions was a Wisdom Receiver he was working on for Intel. The Wisdom Receiver was able to detect

Mark Thomson with one of Wayne Satre's 'Advice To Travellers' outback road-signs, 2017.

only received wisdom, he said, but designed to also screen out artificial intelligence, and 'designed to detect the differences between data, information and knowledge'. Although his projects had little evident utility other than to crack jokes, Thomson didn't appear to be role-playing in the overtly literary or re-enactment ways steampunk tinkerers did. 'I quite like steampunk stuff but don't consider myself part of it as it is just a little mannered,' he said. He said he was more comfortable categorising his genre as 'bushpunk'.

The Wisdom Receiver was designed with dials and bakelite knobs on a pastel steel case. The various-stage prototypes for this and the Hoke range demanded scavenging and close attunement to the typefaces, graphics, hues and sensibilities of earlier eras, and various methods for making the artefacts plausibly 'old'. For these, Thomson used hand-rendered techniques as well as Photoshop and mechanical methods. Still, he didn't call his projects 'art'. He said: 'They're not art, they're entertainment, mainly for me. I love it that people get a laugh out of them. I usually call them props, although since I have to write a whole lot of official museum bumf for insurance purposes,

I've slipped into calling them "objects" because that's the language cultural bureaucrats use.'

Cultural bureaucrats occupied a special place of disdain in Thomson's thinking. When I used the word *curated*, he told me it's a word 'I loathe' for its 'connotations of rarefied, self-conscious and mannered artiness'. When I asked about his *practice* he mocked the word, telling me it was the officious language of bureaucrats and academics, not a word used by tinkerers. He told me that the accepted division between creative and non-creative production is 'an artificial one.' But now, calling his tinkered projects *objects*, he'd subsumed arts-administration language into his dealings. Miles Park's definition of tinkering as 'opportunistic ... very much about the top-down meeting bottom-up' was taking on a new salience here.

The practice of tinkering has always been with us, but a certain self-consciousness hasn't. During my research, tinkering became increasingly publicised, organised, broadcast and institutionalised in Australia and globally. Bit by bit, Thomson's private tinkering values became eroded by the more regulated values of bureaucrats and public institutions. To Thomson, tinkering was largely play: autonomous, voluntary, free, improvising, untethered from time and largely for its own sake (rather than strictly outcome-driven). But through Hoke, as Thomson started playing with broader cultural conceits, this freedom was by increments lost. He welcomed Hoke's public successes, but they came with certain professional impositions. One was a kind of art-world image-management that's antithetical to the tinkering mindset. During my research, Hoke metastasised on a national scale: Thomson had a commission by the ABC to produce a script for *The Lost Tools of Henry Hoke*, a 21-part documentary series funded by the SA Film Corporation and Screen Australia. The Hoke enterprise had become formalised.

Clearly, this bothered Thomson. In a heatwave during the summer of 2012, he and I met up in a Melbourne pub to discuss how Henry Hoke was travelling. He felt uneasy about the way some of the conventions of television were too fixed, too heavy-handed to adequately portray Hoke. 'Transferring this stuff into the popular conventions of what makes good television is extremely difficult,' he told me. He said many of Hoke's trade-culture sensibilities are not jokes *per se*. 'They're not funny at all,' he said, adding that the humour was more implicit, understated and more *material* than conventional idioms can portray. And there was the problem of creative control. More than wishing to *retain* creative control, Thomson wished to *unleash* it – to give licence to Hoke's reported public sightings, memories and artefacts. 'That's the best part of the Henry Hoke thing – that people feel they can tell their own stories. I really want it to go forth in the world and, as you say, metastasise. I want to lose control of it in the best possible way.'

But there were copyright issues, and other television industry conventions that didn't align with the tinkerer's expansive and polymath mindset. Being writer and also creator of sets and props, Thomson frequently found himself fielding questions about what he was doing on the film set. Even within a creative industry such as filmmaking, the idea of the author had strict parameters. The Hoke tinkering had grown from impulsive autonomous domestic play into a role of a more considered custodianship of character (or enterprise) that the television industry, unattuned to this kind of prank vernacular and its tinkering values, didn't entirely understand. Tinkering is a generalist practice, but, like the Hoke enterprise, the practice is invariably about particularities. (Thomson once described tinkering as 'an intense focus on a small closed world'.) More often than not, the tinkered project involves shared insider sensibilities that are attuned to material and circumstantial specificities. While much of Hoke's creation had been collaborative (some ideas and contraptions were built with other tinkerers, or built upon trade-culture references), when top-down production regimes were imposed, certain sensibilities were lost in translation.

Years earlier, Thomson had driven me to visit the Hoke artefacts where they were on display at South Australia's Murray Bridge Regional Gallery. He said he doubted whether many of the sorts of people who would appreciate Hoke would set foot in a fine art gallery. There, unlike the farmers' field day, they were displayed within white walls, under glass, on plinths, accompanied by didactic panels that treated Hoke's history as fact. Only the loud clanks and whirrs of the Random Excuse Generator and the Quack of Doom disrupted the gallery's formal serenity. In Chris Block's story (Chapter 3), Block and Thomson hauled the Random Excuse Generator to a farmers' field day, where it prompted some delight but some tut-tutting. But when you visit these kinds of objects in a public gallery space, you might approach them with a certain reverence; you might expect them to operate within a certain field of signifying practices within theoretically informed, worldly discourses. Thomson's products, on the other hand, operate at the level of mockumentary. More blunt lampoon than sharp satire, they're at once understated and heavy-handed; at once inventive and *bloody useless*. Despite the fine artistry involved in their fabrication, despite their skilled articulation of cultural memories, they remain clunky and unworldly in a high-art sense.

And there's no artist's statement or biography accompanying them. In the gallery visitors' book, Hoke received a range of delighted comments, but the gallery also received complaints, including one that the messages generated by the Random Excuse Generator (largely well-worn political platitudes) were offensive. From a free-range tinkering exercise in an Adelaide home to a national touring exhibition, a book, a field-day demonstration, a Science Week exhibit, an Intel commission and a television series, Henry Hoke weathered many problems of projecting insular domestic tinkering to public categories and gazes. These problems directly accord with our cultural expectations of 'consequentiality' imposed upon our material products. By what benchmarks are the products of people's tinkering to be understood by the rest of us?

Circumstance and place suggest one answer. At a Maker Faire or Burning Man event, these products function as shared, sanctioned and celebrated expressions of making for-its-own-sake. In the context of the farmers' field day, we might impose traditional engineering (or utilitarian) expectations. In a gallery, these same products are freed from their 'working' obligations but instead must express meaning, or certain aesthetic concerns. If the finished product of tinkering is all that matters, then it might be fair to understand these products as existing in a realm somewhere between art and engineering – and as such, simply as repositories of the conditions and values that gave rise to them.

But by now this book has shown that the products of home tinkering aren't all that matter. For tinkerers, tinkering with the material can be a playful way of tapping into the immaterial, and thus the 'inconsequential'. As much as the tinkerer is a cultural interloper, and as much as tinkering – as pleasure or utility – can be seen as a form of culture-jamming, it is also tinkering's *inconsequentiality* that gives it such allure for the practitioner. ('They're not art, they're entertainment, mainly for me.')

That people can pursue their private, dilettantish, inconsequential explorations, fantasies and adventures, without state or market interference, without purpose or externally-imposed temporal and life regimes – this is an important measure of personal freedom and autonomy. Tinkering, as Robert Dessaix has written about amateur production, need not *accomplish* anything. Oriented towards the home, tinkering is a practice that distances the tinkerer from public life in ways that are spatial as well as figurative. In a 2010 *Popular Mechanics* editorial, 'Why Tinkering Is Always Time Well Spent', editor Tyghe Trimble asks readers: 'As a true handyman and tinkerer, do you have to create something that is useful?' He then writes: 'the answer is no. There is value in doing something in and of itself.' Simply to play with things and ideas, simply to absorb yourself into your own tinkerly impulses – tinkering might be inconsequential, and hence an unfettered expression of humanity.

Thomson's relationship with exhibitionary 'sumps' would become even more complicated. Having returned from the San Francisco Bay Area Maker Faire, he was rounding up other makers. In an email, he wrote to his foreign colleagues: 'I am very busy trying to get an Adelaide Maker Faire up in the next month or two.' He wrote that doing so 'is proving difficult – Australian makers and artisans do not feel any obligation to go out and sell themselves, seeing it as behavior on a par with real estate salesmen, used car dealers, politicians and sundry touts ... I suspect this reticence is a crucial difference between the US and the outliers of the British Empire.' (In fact, few of my research participants showed their wares on social media accounts.)

In the meantime, the Henry Hoke enterprise had become increasingly consequential, and Thomson found himself refusing requests to sit in zany poses with Hoke contraptions for newspaper photographers. Meanwhile, Intel had given him a MakerBot 3D printer to see what he might do with it. He decided to share it, housing it in Adelaide's new FabLab. Although Intel's financial patronage had bestowed upon Thomson certain freedoms (he'd quit his government job on receipt of the grant), in small subtle ways it also encouraged a certain pressure and self-consciousness. He said the Intel arrangement was 'very informal but that's not surprising as Silicon Valley funds all sorts of unusual projects. Many science fiction writers get some sort of support from the likes of Intel for instance, which is an acknowledgement of the need to keep imaginative daydreamers linked to the development of science and technology.'

To me, Thomson's own tinkering had produced material forms of science-fiction, just as steampunk tinkerers had, and I guessed this was how Intel saw it. I asked him about an incident at the Adelaide Mini Maker Faire. Attending the Faire was Theresa Famularo, then a director of Sydney's Vivid Festival, who appeared to know Thomson. She approached him, nodded towards his Toroidal Vortex Smoke Ring Generator, saying: 'Mark, *why*?' It was a question that, to me, seemed to invite an artistic statement for a curatorial gaze. Why would you build a rusty corrugated iron tank backed with a rubber membrane that puffs out smoke-rings when gonged with a felt mallet? German sociologist Theodor Adorno, in *Aesthetic Theory* (1970), loftily described the unintegrated, art-for-art's sake motivation as 'infantile tinkering'. I wondered whether Famularo saw this Maker-Faire exhibit as art-for-art's sake, or whether her question was just banter. There was general cheery chitchat between Thomson and Famularo, and he evidently treated the question as rhetorical – he didn't venture a specific answer. Later, he told me: 'People *all the time* ask me: Why are you doing this?' He said that kids never ask such questions, because they automatically get the concept of play.

While serving on the Adelaide Mini Maker Faire organising committee, Thomson experienced more mismatches between tinkerly expectations and those of creative institutions. First held in 2006, the Maker Faire was a spin-off from the US-based *Make*

Thomson and Maker Faire spectators with the mini version of Thomson's Toroidal Vortex Smoke Ring Generator, built with fellow tinkerer Phil Gerner, at the Adelaide Maker Faire 2016. (Source: Adelaide Maker Faire).

magazine, created by Dale Dougherty, a serious proponent of maker culture (whom Thomson regards as 'genuinely altruistic'). A free and ostensibly non-commercial event, Maker Faire has spawned licensed events internationally and continues to proliferate. It's now, according to official promotion, the 'World's Largest Show (and Tell) festival – a family-friendly showcase of invention, creativity and resourcefulness, and a celebration of the Maker movement. It's a place where people show what they are making, and share what they are learning.'

The Maker Faire phenomenon exemplifies many of what academics like to call *turns*. In the scholar's space, there are linguistic and cultural turns, literature turns and historiography turns, ontological turns, moral turns, digital turns and transnational turns. There was the material turn in the humanities and social sciences; there was the writing-culture turn in critical anthropology; there are spatial turns in architecture, design, philosophy and history. There's some scholarly quarrelling about 'turn talk' and the 'narrative of turns', but academics tend to regard turns as moments in recent history or postmodern breaks from tradition, when former theories and methods (often positivist ones) become outmoded and new ones (often constructivist) emerge. Turns are always rancorously contested, and their arguments spill outside the ivory tower and seep into media and political fulminations (hence, the widely-debated history wars, culture wars and science wars of the 1990s and beyond). Many turns peaked between the 1960s and the 1990s, but in museum and curatorial fields there's ongoing discussion of a turn away from art-institutional-led exhibitions. This turn steers instead towards outsider institutional forms, such as the branded or themed travelling biennial. Maker Faires are one of these forms. In museum practice, there's also a current curatorial turn that casts its gaze away from the obscure, and towards the everyday. Again, Maker Faire fits this bill. Another development is a move toward less contemplative and theoretic curatorial practice, and more activist practice. Maker Faires offer this. And there are other turns, yet unnamed. Exhibitionary practice is moving away from artistic and curatorial heroicism, and instead 'engaging viewers as cocurators', according to curator Terry Smith. Some exhibition spaces are moving beyond the 'highly-static human-to-object relation' and promoting person-to-person interactions. Again, as we see in Maker Faires. So let me put it to the turn-sayers and the jargon-coiners: the tinkering turn is upon us.

As I write, 'The Museum of Everything' is open at Tasmania's MONA, with a stated aim to advocate for 'the work of ordinary people' and 'art that falls outside the confines of the art world', made 'not for you or me, but for themselves'. The tinkering turn also manifests in the advent of MEND*RS. MEND*RS is a conscious way of tinkering with, disrupting and reforming public space and discussion. Several scholars have documented organised movements of yarn-bombing, craftivism and stitch 'n bitch, but 'cultures of mending' scholarship isn't yet precisely defined and demarcated. It has a nascent basis not just among crafters, but in scholarly action-research. The 2012 advent of Lancaster University's MEND*RS project, an online 'Mending Research Symposium', set out to 'devise a critical agenda for mending research and to disseminate and promote mending practices and enterprises on a local, national and ultimately global level.' MEND*RS carries a manifesto for social and political change (pictured in Chapter 2) that's strikingly similar to the Makers' Bill of Rights and the mission of maker movement proponents, aiming 'to collectively bring about "The Age of Mending".' Unlike the more Silicon Valley-aligned maker movement proponents, MEND*RS is scholarly, overwhelmingly female, focused on traditionally female practices, and concerned with aesthetic, historic and political research.

At the time I was finishing my research, a multidisciplinary 'Cultures of Mending' workshop was organised by the University of Chester's Rebecca Collins, whose work as a human geographer investigates many tinkering themes, including deep engagement with material objects, and wellbeing fostered by manual competence. Collins was also interested in the public and political implications of mending, and 'the political potential of mending as a critique of capitalist society'. To the uninitiated, the small act of darning or sewing up jeans might seem modest and inconsequential, but fashion historian Anna König is among many scholars who view mending as an activity that 'should not be positioned as a trivial activity' and that 'is profoundly important in environmental, social and economic terms, and not just as a metaphor, but as a democratic action that is as materially tangible as it is empowering'. Mending is a kind of tinkering that involves the diversion of objects from their pre-ordained or designated paths, and it very often assigns value to the spoils of other industries. Unlike *making*, but decidedly like *tinkering*, *mending* is a difficult idea to exhibit publicly in forums outside fixer cafés – but its practitioners are people who reorient our collective knowledge, and form continually evolving hubs of arguments with materiality and power.

Adelaide's Mini Maker Faire, said Thomson, was organised by an 'ad hoc committee' that included representatives from ANAT (Australian Network for Art and Technology), DFEEST (South Australian Department of Further Education, Employment, Science and Technology) and FabLab Adelaide. Maker Faires run on voluntary labour, but sometimes with help from sponsorship and trading exhibits. To recruit exhibitors, the Adelaide committee 'put out a call for makers and accepted most of them'. Much of the recruitment took place through social media, and some exhibitors were hi-tech companies such as 3D printer developers. Others were individual makers with not a mili-digit of social media presence – Thomson personally recruited people like tinsmith John Yard, who doesn't use telecommunications beyond a landline telephone in one of the outhouses on his Goolwa property, which he answers only occasionally.

I asked Thomson how the Maker Faire was curated. He said the event wasn't *curated* (he hated that word) but he told me in an email: 'They will probably have to be increasingly 'curated'... We had to keep out the dream catcher types and the sort of people who inhabit craft fairs with imported crap. There could be a few difficult choices to make with that sort of thing. You could keep it to the usual suspects – the hackerspace/FabLab/groovy young digital literati or move out to the people like the blacksmiths association and everything in between. Not easy to manage fairly.'

The Melbourne Mini Maker Faire did not itself emphasise fixing, but enlisted fixers for the forthcoming Sustainable Living Festival.

So there were certain issues of branding and status. The positioning of exhibits in any public spacc is inevitably political and strategic, and also determined by the site. Held on various floor-foyers and nooks of the multi-storey Adelaide College of the Arts building, the Mini Maker Faire was a panorama of demonstrations, stalls, exhibits, robotics and electronics workshops, as well as brooch-making, bookbinding, sock-puppeting, moulding and casting, and seminars by celebrity speakers. Celebrity makers and popular objects (*The Curiosity Show*, Daleks) were positioned at the ground floor entrance; more white-collar robotics and high-tech workshops were positioned in labs above handcrafts; and down the side in the basement carpark and on the outside lawn were dirtier, blue-collar projects: tinsmithing, blacksmithing, cars, bikes, the smoke generator.

According to ANAT, there were four thousand attendees to this first Faire. Thomson said he planned to divest from future involvement (he didn't), but that the event was 'an all-round big success' in terms of attendance and exhibits, many of which – the Dalek impersonator technology, a light model plane, and Thomson's own Toroidal Vortex Smoke Ring Generator – went on to feature in National Science Week demonstrations.

The previous year's Melbourne Mini Maker Faire wasn't organised with arts institutions, but by Melbourne Hackerspace and Swinburne University. It largely featured new tech-science (including advanced biotech), though traditional crafts such as felting and dressmaking were represented to a small extent. Thomson had told me: 'I have no doubt that Australians are more focused on fixing than Americans', but meeting him there we both remarked that the Melbourne Faire didn't emphasise fixing (it recruited people for fixing workshops to be held later at the Sustainable Living Festival). The emphasis was instead on relatively new fabrication technologies such as 3D printers. Repair, maintenance and adaptation received comparatively little attention. By contrast, the 2009 Maker Faire Africa focused almost exclusively on recycled, upcycled, makeshift, adapted, fixed and folk technologies.

So the monocultural framing of a global 'making culture' seems to work more as a marketing tool than an explanatory one. Thomson told me that framing maker culture as 'science' is often more successful than framing it within an arts institutional context. During my research, he took up a position as speechwriter for another South Australian minister, a role which also included advice. The government, he said, 'has discovered the importance of STEM [Science, Technology, Engineering, and Maths] in a big way and [has] thus written it into a whole raft of fine-sounding policies, strategies, masterplans etcetera'. But he said a problem remains – that of people's disengagement from production of technologies. He described this as 'a long-term cultural problem: technology is now so complex, reliable, cheap and pervasive that it is no longer necessary to understand how or why it works. People just use it.' In an email, he wrote:

> As the Swiss writer Max Frisch said, 'Technology is the knack of so arranging the world that we don't have to experience it.' Thus technology has reverted to a magic black box, which only some people will understand … if they've got STEM knowledge. Altering that cultural shift is well beyond the ambit of what modern short-term focused governments can do to change. But because STEM knowledge is the foundation of our prosperity, government can only keep repeating STEM, STEM, STEM as a mantra whose repetition will somehow cause wholesale attitudinal change in young people. It's not good enough.

If the values of everyday tinkering can't be formally tallied; if tinkerers can't be an organised collective; if our cultural and educational institutions can't adequately recruit and represent these people – then how can we possibly name them, or imagine them as a tribe? Already, the stories in this book suggest that one way to conceptualise Australian tinkerers is to understand them as *good workers*. Recent cultural studies scholarship offers some other useful ways.

In *Locating Cultural Work* (2012), Australian craft scholar Susan Luckman describes what she calls 'the slow creative industries movement', a category of people who see meaningful and sustaining work as of central importance to personal ethics and identity. These people equate 'the good life' with 'the slow life'. Luckman also points out – along with sociologist David Gauntlett – that such apparent 'movements' are cyclical. Many scholars have understood current maker culture as a contemporary reiteration of the values of arts and crafts movements (or convergence of movements), widely understood as a wilful return to artisanal values in an industrial age.

Yet this doesn't account for those who've always tinkered, heedless of cultural fashion. And in identifying a contemporary 'slow creative industries movement', Luckman also qualifies that this 'is to imply a far greater level of consciousness and collective organisation' than is apparent in her own research of creative workers. So a useful way to think about tinkerers is as a category of public, rather than as a movement. In *Publics and Counterpublics* (2005), social theorist Michael Warner describes a public as 'a space of discourse organised by nothing other than discourse itself'. To Warner, a public can't be defined independently of the rhetoric through which it's imagined. Anthropologist Christopher Kelty, too, understands a public as a social imaginary.

Tinkerers can be understood as a counterpublic, or within a framework that Kelty describes as a 'recursive public', a public that's actively 'concerned with the material and practical maintenance and modification of the technical, legal, practical, and conceptual means of its own existence'. A recursive public is 'independent of other forms of constituted power'. This public is 'capable of speaking to existing forms of power through the production of actually existing alternatives'.

In other words, recursive publics – like tinkerers – are continually evolving hubs of arguments with materiality and power. Warner, too, describes this 'reflexive circulation' within a public. Recursive publics differ from more formal public groupings because of their 'focus on the radical technological modifiability of their own terms of existence'. What Kelty means is that recursive publics are themselves mutable even as they mutate their material and political resources. When humanity tinkers with things, those things tinker with humanity.

For everyday tinkerers, this reorientation of power is embodied in material things, but also in the practices, mindsets and social engagements made possible (or enabled) by those material things. Or the narratives attending those material things. So the tinkerer who accesses social media or a *Make* magazine to learn how to fabricate a tricky component might unconsciously engage with discourses surrounding contemporary fabricating and fixing. They may apply and build upon that knowledge and share it with peers. Likewise, the concept of recursive public 'bypass[es] the dichotomy between ideas and material practice'. This resonates with Sterne's theory of *techne* as both creation and contingency.

When Kelty thought up his recursive public theory, he intended it as an explanatory framework for software – not hardware – developments and communities: creative commons, open source and open access movements. Yet, to Kelty, software and code are themselves material forms. So it's not such a conceptual leap to apply his framework to material tinkering with hardware forms. Still, many open-source movements are concerned with infrastructure of some sort, and so have some loose level of organisation and *explicit* ideological *intent*: not always the case with everyday tinkerers. Tinkerers aren't a neat ideological collective. And nor is there a sharp distinction between the individual and organised concerns of tinkering. There is not a clear line between the tinkering world's public and private spheres, either. This is especially so at a time when online global connection has effectively turned our private (civil) homes into public (civic) spaces. In the tinkerer's space, where patent blueprints are copied, houses are illegally wired, components and patterns are traded, software and hardware are routinely tampered with against the terms of their licences – in these events, civil disobedience can blur with more civic disobedience.

This is why Kelty's recursive public model works so well in the tinkering subterrain, where there's no clear distinction between global, local, public and personal. Members of recursive publics don't necessarily set out on their activities with conscious, ideological, political or entrepreneurial *intent*. Instead, they might form or confirm political and ideological positions through their *direct engagement* with creative practices. Unlike the 'public sphere' as a generalising category, the recursive public is focused primarily on *production*, with social transformation as one of production's impacts. Recursive publics can be entanglements of diverse and dispersed people whose *very making practices* can form 'a kind of argument, for a specific kind of order: they argue *about* technology, but they also argue *through* it'. In other words, legal and organisational mechanisms are challenged through material production and its processes.

Three years before the Adelaide Maker Faire, I interviewed a tinkerer named John Bennett, a bloke with whom Thomson had once worked. A straight-talking inventor who once ran a successful engineering firm, Bennet told me stories about the green commercial buildings he'd designed. He'd also built race-bikes that earned considerable fame in racing circles. Bennett described material tinkering as subversive: a way to 'make you look at the world in a whole different way, a critical way [that] moves you from a lesser truth to a greater truth'. Although this sentiment precisely iterates maker movement rhetoric, Bennett had almost no web presence and didn't engage by email. At Thomson's urging, he attended Melbourne's Mini Maker Faire in 2012, but he didn't express interest in exhibiting his own inventions; nor was he excited about the potential of 3D printers. What are the differences, I asked Thomson, between the sorts of people exhibiting at the Faires and the John Bennett types? He wrote in an email:

> It's useful that you've mentioned John Bennett because he has a favourite expression: 'less yap-yap and more tap-tap', that is, less talking about doing or making and more actual doing or making. It might come from a workshop setting where workers are slacking on the job but it also encapsulates an important attitude – that doing the work is more important than big-noting what you might or could or have done. And some people are simply uninterested in the 'show and tell' aspect of Maker Faire … the prospect of sitting around all day spruiking something you've done is frankly boring.

Engaging with discursive (official) and applied modes of tinkering, Thomson himself was equally yap-yap and tap-tap. But despite the obvious successes, vitality and goodwill of the Adelaide Mini Maker Faire, Thomson said he wouldn't be part of future Faires organised with institutions that are 'turning to makers to bring the glamour of the crowdfunded start-up, the heady glamour of Silicon Valley and the cutting edge chicness of the 3D printer to their own stodgy ways'.

This isn't just an insider perspective. Craft scholars including Anthea Black and Nicole Burisch describe an 'institutional appropriation' – an opportunistic method by which institutions gain economic, intellectual, political or cultural kudos by exhibiting cultures that 'make the institution seem visionary and fresh'. This helps explain why the meaning of *hacker* has been incorporated, inverted and ultimately diffused in government rhetoric (see Chapter 2). In her essay about the culture-jamming *Brand.New* exhibition at Britain's Victoria and Albert Museum, museum studies scholar Kirsty Robertson wrote about the ways that the exhibition 'was deeply implicated in a re-branding of the museum itself'. Expressions of resistance can be diffused when co-opted into public spectacles. In their essay 'Craft Hard, Die Free', Black and Burisch write that when material expressions of resistance are used to engage and develop new audiences and markets, 'it shifts the value of the work away from the original practices'. In other words, the act of contextual abstraction can obscure the very essence of tinkering.

Thomson told me he preferred the more irreverent ground-level Macclesfield Gravity Festival (now known as Gravfest) to Maker Faires. Gravfest, he said, is 'less earnest. It's more fun. It doesn't take itself so seriously.' As people set up stalls and road-blocks, the town's excited charge was palpable. A couple of my interviewees had invented carts and G-bikes for Gravfest, a high-speed downhill race between DIY carts on a steep road through the idyllic pastoral South Australian town of Macclesfield. They'd spent many months before the event designing and crafting their projects in the kind of unpaid production regimes that some ethnographers and sociologists have interpreted as 'self-exploiting'.

Although Gravfest was organised and regulated, it had an informality and homespun festive vibe that the Maker Faire, with its scale, its scope, its program of international and celebrity speakers and its high-tech events, didn't. Nor was Gravfest promoted in the official mission-statement style that had framed the Maker Faire. Instead, when I contacted Gravfest organiser Fred Keal, like Thomson he had a subtle register of blokey, anti-intellectual, deadpan lampoon. When I mentioned I was working on a PhD, he told me ABC radio had just interviewed him, and he had 'mentioned I have a PhD. The interviewer seemed surprised until I went on to explain it stands for the "post hole digger". I used it to drill into the deep vein of daftness that runs under Macclesfield that's believed to be the cause of our prime gravity.'

Keal explained that Macclesfield's prime gravity caused its residents to sometimes lose balance and fall forward just outside the pub door as they left. He told me he had 'always been a tinkerer' because he gets 'immense satisfaction from the process of producing something from nothing or making something work that didn't'. Although he works as a state police prosecutor, Keal, a recent migrant from the UK, expressed contempt for Australia's 'risk avoiding over-governed society' and his comments suggested a beating-of-the-system pride that characterises all the tinkerers I interviewed. Still, Gravfest was far from a lawless event. 'Gravfest looks like a bottom-up affair,' he said, 'but it isn't entirely as it seems, as my experience in putting together police operations has made me a bit autocratic … I like the phrase "a camel is a horse designed by a committee".'

Keal said he has 'a deep mistrust of committees' and 'people who talk and don't do' (redolent of John Bennett's 'less yap-yap and more tap-tap'). In contrast to Thomson's experience with a committee-led approach, Keal evaded the frustrations of top-down managerial approaches to the risk management Gravfest had to negotiate – road closures, public liability insurance, and health and safety – by exploiting the tinkering mindset. 'The risk of ending up with a camel is huge if you let eccentric blokes mess with your plans. Tell them the build spec is set and they will accept it. They will try and push the boundaries, which is part of the fun so … ensure there are some to push at.'

In her 2011 *Time* magazine article 'In Praise of Tinkering: How the Decline in Technical Know-how is Making Us Think Less', science journalist Annie Murphy Paul wrote that tinkering 'involves a loose process of trying things out, seeing what happens, reflecting and evaluating, and trying again'. This was manifest in the

Gravfest approach. To arrange official council support and approval, which included a comprehensive risk management plan, Keal said: 'I approached the newly appointed Mount Barker District Council tourism manager and rolled him down the hill in my car in neutral to show it reaching 85 kilometres per hour. This is where the wild-eyed loon bit came in because he could see I had the vision and drive to make this happen and that could never have been achieved in a letter. It's all about people; find this person in your area and take them for coffee.' Cultivating contact in this way was 'vital to the whole operation'. To measure Gravfest participants' speed, Keal had enlisted local police. Local voluntary community organisations that routinely deal with insurance, permits and health and safety issues – Lion's Club, St John's Ambulance – were more strategically useful than more formal state institutions such as schools. He told me in an email:

> The kind of people who become head teachers don't do risk. They lay awake at night worrying what might happen to the children the next day. They will not commit to such an event and might be at the root of our risk aversion problems … Middle aged men who should know better will be queuing up to risk their lives on your hill and they will bring money and people to watch.

At Gravfest, some crafts were constructed in situ. Participants brought tools along to assemble and disassemble their crafts and to load them on and off trailers. Girls and women were included, but mostly as spectators and supporters. As Keal noted, Gravfest participants were mainly 'middle aged men who should know better'.

Adrian Matthews's wooden bike outside his shed the evening before Gravfest; the finished bike at Gravfest.

One of these was Adrian Matthews, whom I interviewed the evening before. He hadn't yet finished his wooden G-bike constructed from an upcycled wine barrel (its design had yet to conform with race regulations). At the eleventh hour, it didn't yet have a seat, handlebars, foot-stirrups or bracing. On the day of Gravfest he told me he'd been up most of the night, and the paint had barely dried by the time the race was up and running (he and his son won a couple of categories).

The event was festive and thrilling, clearly to everyone. After the races, prizes were awarded, participants and families stuck around to admire each other's handiwork, discussing the value of various suspensions, brakes, chassis, steering mechanisms, bodies, finishes and materials (fibreglass, ply, cardboard, plastic). People were disassembling crafts and stacking them on trailers, and kids gathered around these, pointing. The milling and banter went on until after the Gravfest stalls – food, produce, craft, tools – were packed away and sunset loomed, and people wandered to their cars or to the pub.

Keal expressed contempt for Australia's 'risk avoiding overgoverned society', reminding me of Thomson, who complained of cultural bureaucrats who are unable to deal with risk. 'There's all this earnest stuff about innovation,' said Thomson, 'but complete institutional paralysis when it comes to risk-taking'. Risk-averse culture, he said, 'has a terrible habit of killing off fun and there were a number of things [in the Adelaide Maker Faire] that didn't happen because of these concerns about possible consequences of possible hazards'.

But in April 2014, when the annual success of Gravfest was attracting participants from around the country, Keal experienced a serious consequence of holding a hazardous public event. A cart in the Open Kart category veered out of control and into the crowd, resulting in two severe spectator injuries and minor participant injuries. A month later, in an email to me describing his distress, Keal wrote of his 'shock and regret that my idea had caused such mayhem' and of how 'I was convinced that it [Gravfest] could not carry on.' But then:

> The grace of the most seriously injured person has been an example to us all as she is 75 years old and still in hospital with a broken leg and elbow but continues to say it was an unfortunate accident at a splendid community event. She has made it clear that the event must continue.

And, of the other injured spectator:

> When I visited the 69 year old lady who suffered bruising in the accident her first question was to ask if I and my family were OK because she had heard how hard we had taken it. That was pretty impressive from someone who had be hit by an out of control billy cart and left me speechless.

From Gravfest's Facebook page it was clear that the community, deeply vested in the event, was eager to find solutions to make Gravfest work, women and men alike. When I asked Keal about this, he wrote of a 'remarkable' and 'humbling' response from 'a fantastic community who clearly want the Gravity Festival to carry on':

> the community has quietly stood behind me in a way I could never have dreamed of. The accident took the event out of my hands and allowed it to be embraced by the wider community … Gravfest

> was brash and dangerous and wild but it was only when it was wounded that people were able to step forward to care for it.

Sponsors and council, he wrote, hadn't cancelled their support for the event. Instead they 'stepped forward with ideas and support for the future'. Keal said:

> There will always be an element of danger and we all know that's why half the spectators are there but the best plan is to keep the carts and the crowd apart at all times. Clearly there is a need for better barriers and that will be in place … the extra cost of improved safety will have to be met but we have resources within the community so it will be covered.

The difference between the risk approaches of Gravfest and the Adelaide Maker Faire may be that the former operates as a culture and the latter as an institution. In *The Risk Society and Beyond*, sociologist Scott Lash differentiates between risk society (a concept introduced by Ulrich Beck as 'a systematic way of dealing with hazards and insecurities') and risk culture. Institutions tend to deal with risk according to norms; whereas cultures deal with risk according to values. Norms, asserts Lash, are procedural and social; whereas values are substantive and cultural. Values are transferred between generations and groups, between people and things. Despite their boundary-crossing posturings, cultural institutions can't be fully modern in their approaches to risk. To Lash, risk is central to innovation.

In contrast to Gravfest, the larger-scale Maker Faire, according to Thomson, had a preemptive approach to risk that took 'the form of a kind of self-censorship or prohibition'. He said administrators managing the Faire presented 'any number of reasons found why fire would not have any role', whereas 'in the US event there were some truly spectacular balls of fire sent up into the sky, often as the result of an 8 year old child pressing a button. They seemed to find ways of managing such things in the US.' Over there, he said:

> personal free will enters the legal argument a lot more whereas here there are considerable restrictions in terms of duty of care … a legal and insurance minefield. In some ways I completely understand that institutions simply throw up their hands and walk away from what would otherwise be simple and actually rather low risk activity.

He said that accepting individual risk is a condition of ticket-sale in the US.

In Australia, Thomson couldn't exhibit a bike-powered merry-go-round. He said there was 'a huge queue' for one in the San Francisco Fair. 'I asked the blokes who were running it if they minded if I pinched the idea for Australia and they said go for it. It didn't happen.' In an email, he explained he would have required 'engineering certification (expensive), which meant that it would have to be built from all new parts (expensive) for certainty of structural integrity, have a complex maintenance schedule, training for supervisors, regular structural checks, safe clean helmets for all users (all expensive)'. He said it 'went from costing maybe $500 to well above $10,000. It wasn't going to happen. And this was a very safe version of an activity that, on any given day, at least half a million Australians young and old do without thinking.'

Thomson later told me that he ended up building the forbidden contraption to his own design in Melbourne, with local government money. South Australia, he said, 'has very tough laws about fairground attraction'. But he said the organisers of Science Week managed risk more creatively than creative institutions. He explained

Above: The US rotary bike contraption designed like a Hill's Hoist that couldn't feature at Australian Maker Faires. Photo: Mark Thomson.

Left: Carts at Gravfest, Macclesfield, South Australia.

that exhibitors must ensure against 'pinch points' that are potentially hazardous. He said arts institutions foreclose opportunities to exhibit at committee-planning stage if a pinch-point is anticipated. Some science institutions approve exhibits first and then deal with pinch-points by retrofitting safety solutions on to already accepted exhibits. The National Science and Technology Centre approved Henry Hoke exhibits and:

> After [the exhibit] was all set up various people from Questacon toured and looked at everything to ask the question: 'what could go wrong?' But to ask the question, they had to find a solution. Everything [potentially hazardous] got fixed. This is so rare – it was fantastic to be amongst a bunch of people whose approach to risk isn't: 'It's risky, so let's not do it.'

Gravfest, too, accommodated dangers as a culture, in ways that the institutionalised Mini Maker Faire couldn't. The Maker Faire was spatially and institutionally organised, mapping out in a one-way flow of information which, though engaging and experiential (learning by doing), was also instructive and controlled. At the Faire, there was unequal knowledge-exchange between the instructor-exhibitors and attendees. Attendees weren't equally vested in the event, and ownership of hazards wouldn't be shared with spectators. On the other hand, Gravfest's exchange of knowledge was peer-to-peer, mutual, informal and experimental – and Gravfest materialised in a site of existing local memory and habitus. Unlike the Maker Faire, it didn't decontextualise or abstract its exhibits. Once they had raced, exhibits were parked and propped ad hoc on the roadside where their pinch-point hazards were left to community negotiation, as attendees probed, police watched and children climbed.

10

Reform

Australian prisons are populated with ingenious objects covertly made – tattoo machines made from CD players, ink made from under-table metal scrapings, weapons made of whittled plastic cups, soap and matchsticks, syringes crafted out of biros. Some prison administrators maintain *wunderkammers* of such things. Covert tinkering also features in conditions of war and oppression everywhere. At the Củ Chi tunnels to the north west of Ho Chi Minh City, tourists are confronted with the ingenious, low-tech weapons improvised by people on the ground out of the remains of their poisoned forests – Punji stakes, camouflaged snake-pits, 'tiger traps', 'keepsake, lose hand' traps – that blunted the impact of US military-industrial might during the American war in Vietnam.

All these products are foils to impositions of power and authority. In Australia, as punitive laws around public participation and protest tighten, tinkerers invent new material forms of civil disobedience. One of the most productive tinkerers I've met is Brisbane-born robotics artist Michael Candy. At the time I met him, he'd just finished a project in collaboration with an activist outfit called Dirty Work. Donning some high-vis gear, Dirty Workers installed Candy's faux security cameras in Brisbane at the G20 Summit in November 2014, despite a strong police presence on the site – a state force directed against its citizens' ability to protest. The 'security cameras' were in fact projectors made by Candy, which he'd designed to be activated from designated mobile phones offsite. On activation, they projected anti-G20 message on the footpaths and surrounding buildings ('G20: Corporate Plunder', 'G20: Business as Usual', and others). Undetected for three days, the projectors were finally removed by police. Perhaps not wishing to draw attention to the incident, police didn't arrest or investigate anyone, but Dirty Work had filmed the devices' installation, ensuring widespread media coverage. 'They've got guns, batons, mace, sonic cannons, water cannons, armoured bearcats and drones,' said Dirty Work. 'We've just got creativity, criticality and love.' Open-source files to help tinkerers build such robotic resistance devices are on Candy's site (http://main.michaelcandy.com/).

Another of Candy's productions was Bitter Bench. This was a public bench he built in response to policy introduced by Brisbane's then-mayor Campbell Newman. The Newman council had removed the city's bus shelter benches, in order to 'clean up' the city and prevent homeless people sleeping on them. So Dirty Workers put on high-vis gear, and 'unofficially' installed Candy's Bitter Bench in a bus shelter in Brisbane's William Street. The bench was set into concrete, but mechanically rigged to tip people off when they sat on it. The Bitter Bench, said Candy, was 'illegally installed as a retaliation to the removal of the shelter'. It accompanied 'a council plaque insinuating the new bench installation was mayor Campbell Newman's penance to the homeless'. The installation included 'a proximity-activated audio system' that played recorded testimonials from the homeless people who had been displaced by the city's new policy. As I finished this book, Candy was working on a 'digital empathy device' – an 'online, citizen-run mapping service that correlates live information on bombings, hazards and attacks in war zones around the world'. This information is relayed through mobile phones to a public statue that Candy fabricated in a hotel room in Paris, which, he said, will be 'mounted guerrilla-style atop the head of the "Marianne" statue, a symbol of the French Republic and égalité, towering above the civilian memorial of the 2015 Paris Attacks'. With each war atrocity transmitted by mobile phone, Candy's statue sheds tears.

Michael Candy's faux 'security cameras' under construction. His website contains an open-source instruction manual on how to build this cellphone-activated device and projector.

Like Adis Hondo's spycam that filmed forbidden footage of events in Washington, and like Mark Thomson's faux objects that bluffed their way into gallery spaces, these are clearly covert responses to impositions of authority. Less obviously, so are the joyous everyday tinkering projects described in this book. These domestic projects help us understand what relatively free people can do with their First World privileges during peacetime. In their everyday negotiations of materiality and power, domestic tinkerers can give us a glimpse of what an engaged, post-industrial, post-work and post-scarcity society can look like. This is especially important at a time when many economists are forecasting a general condition of 30 per cent unemployment in the next decade or two. It's generally accepted that the era of full-time work is nearing its end, in part because of the nascent loss of jobs to automation (driverless cars, self-check-outs, 3D printers, automated booking) and online technologies (in my field, algorithms that write articles; social media content taking over news production).

This book is written at a time when 1.1 million Australians are underemployed; an estimated five million of our existing jobs will be lost in the next decade; more than one in four of us would like to work less; three-day weeks are starting to emerge as workplace norms; and policy makers across the political spectrum are discussing the possibility of a universal basic income (UBI) to prevent an inevitable underclass. Some are considering an economic model in which tax revenue from corporate giants would allow everyone to be paid a non-means-tested UBI that we can supplement if we have paid work. In other words, regardless of our circumstance, we'd get paid to live, and do with our lives what we like.

As I finished writing this book, several new-release books discussing UBIs emerged: Melbourne academic Tim Dunlop's *Why the Future is Workless*; American historian James Livingstone's *No More Work: Why Full Employment Is a Bad Idea*; Dutch journalist Rutger Bregman's *Utopia for Realists;* Sociologist Peter Frase's *Four Futures: Life After Capitalism;* and Irish cultural studies scholar Paul O'Brien's *Universal Basic Income: Pennies from Heaven* – among many others. Collectively, these books present the case for a basic income whether or not we participate in the formal labour market. They show why formal work shouldn't be conflated with good citizenry, and they map out the ways work can become untethered from income. Formal employment as we know it is not the solution, but the problem, according to Livingstone. These books also document a mounting body of evidence that modern economies – even free market ones – can prosper without people being required to toil in the full-time formal labour market. (Many progressive economists like this model's egalitarian potential; many conservative ones like it because it would essentially abolish the welfare state.) The UBI idea isn't new, and nor does it automatically enable an anti-capitalist model. It wouldn't fully redistribute wealth, end wage disparities, or bring instant equality to women, who have always performed unpaid work in the home. But it would lessen these inequalities, and trial studies on sample populations across different economic models (in the US, Finland, Kenya, Namibia, Canada and Netherlands) have surprisingly positive results – most people become *more* entrepreneurial and *more* productive when supported by a basic income. This is in part because an unconditional income gives people the freedom and agency to take capital risks and make genuine choices. (So notions like 'enterprise bargaining' begin to have real meaning. We won't settle for poor pay and labour conditions if our livelihoods don't depend on it.)

More, it turns out that most of us *want* to work. We want to be useful and engaged. We have motivations to work outside of its monetary rewards. As I've documented in this book, public intellectuals including anthropologist David Graeber are mounting public campaigns against 'bullshit jobs' – work that's meaningless and destructive, and jobs that people wouldn't choose if they had real agency. American economist David Korton believes many of us would use more leisure time to engage with our communities and campaign for a better world, and Rutger Bregman's *Utopia For Realists* (2017) supports this assertion with concrete evidence, as well as evidence that societies with fewer working hours enjoy *greater* economic productivity and healthier, happier citizens. But, as I write, there's some discussion about what Australians would do without full-time jobs. Will we be a nation of bludgers? How will our humanity be impacted? Will we finally enjoy the leisure time that automation always promised? What will we do in that time?

Mass-leisure industries tend to be highly consumptive, and they tend to have a hefty environmental impact. An overseas holiday, for example – even with carbon offsets – is hardly justifiable at a time when we've created our own Anthropocene. In terms of carbon output, a single overseas adventure can eclipse a lifetime's effort of riding to work, recycling our rubbish, lowering our food-miles, eating less livestock, using cloth nappies, buying sustainable produce and installing solar panels. Stories in this book demonstrate other possible leisure-time models. They show how, to its practitioners, tinkering figures as adventure – as expansive and educating as the most thrilling travel expedition, even within its insular and everyday focus. In Civitico's story of jewellery-making at home, she 'travelled' along her supply chains, developing a connection with lampwork trades in the Northern Hemisphere: 'People slave over a hot torch flame to produce these from coloured glass rods,' she told me. In John Tucker's story, the fixing of his power-saw generated an expedition of global reach from his desktop, during which he accrued knowledge about the ways component parts are made and mobilised. People become tourists when they tinker.

More, studies show (unsurprisingly) that cultures whose citizens spend fewer hours working in formal jobs also consume less than those with longer formal work-hours. In such cultures, 'work–life balance', a phrase that suggests that each is in opposition, might become meaningless. Because tinkerers' lives aren't fissured between work and leisure, tinkering is labour that isn't toil; voluntary but not leisure. Instead, the tinkering described in this book is a vocation that can sometimes provide income, but always tends to wellbeing and livelihood. It's a moral and economic activity not tallied in formal ways. It doesn't require recouperation. It complicates relations between production and consumption. It doesn't have clock-in-clock-out values; instead it forms a cumulative life-narrative. To sociologist Richard Sennett, good work itself isn't an endgame, but the process of amassing skills and knowledge. Vocation, likewise, isn't simply a productive career, but a habitual 'sustaining narrative' or evolving 'work story'.

The tinkerers I met were people who can build productive lives without exploiting those of other people, nor those of the planet. Although tinkering uses material resources, much of the tinkering described in these pages delays the escalation of material exploitation and climate collapse. During Michael Drinkwater's conversion of an old dishwasher into a working biofuel processor (described in Chapters 4 and 8) he operated under his own rules of engagement, exploited the spoils of other industries, created his own more-sustainable one, challenged material categories and subverted manufacturer intent. By doing all these things, he formed social relationships of trust and informal economic exchange (beer exchanged for used vegetable oil, glycerin by-product exchanged for mechanical repair, and so on). Similar social reciprocities circulated during Gilda Civitico's fabric remnant creations; Adis Hondo's shed-building; Irene Pearce's tank-home; Kate O'Brien's portraits; Chris Block's iSlate; Mark Thomson's ostensibly useless Random Excuse Generator, and so on. All these everyday projects had impacts and trajectories beyond their maker's conscious intent.

Each tinkerer's project seems an insignificant way to change the world at a time when we've discarded an estimated 65 million tonnes of electronic waste, and we're consuming more than 300 million tonnes of plastic, much of it ending up in landfill and oceans. But the aggregate effects of individual action can be significant. The tinkerers in these pages engage in a tacit understanding of the inseparability of human history, material history, 'natural' history and geopolitical action. In the stories of Tucker, Hondo, O'Brien and Thomson, this was especially explicit. These tinkerers express a material riposte to our politicians' conceit that the material world is our servant, rather than our collaborator and benefactor. This is an ahistorical view and a dissonant delusion – one violently ridiculed by anthropogenic climate change, a geophysical backlash against industrialisation's dirty excesses (in free market and non-market societies alike). 'The distinction between human and natural ontologies is at an end,' writes US historian Timothy Lecain, whose book *The Matter of History:*

How Things Create the Past is being released at the same time as this one. Since the Enlightenment, many of us have seen ourselves on a plane above the material world, because our ontology 'was one of creation and self-creation. Humans were subjects, never objects.'

But the self-creation mindset underpins the Australian political discussion, and so the tinkerers I researched tended to believe that our livelihoods are too important to be left to policy makers. A UBI might allow more people to tinker, but for now, at a time when our neoliberal state is junking the social contract as it trashes the planet, many tinkerers are inventing modest ways to reclaim their senses of humanity and agency, and in doing so aid the humanity and agency of others. This isn't a substitute for direct political action in the public sphere – it's an augmentation. Most tinkerers I interviewed were engaged with the political as well as the material sphere; the relationship between the two being recursive.

Repair, for example, is a growing practice of collective global mobilisation and local initiatives – community tool libraries, repair cafés, Hackerspaces, iFixit.com, Restarter parties – that seek alternative relationships with First World production. (As do community produce-swaps, craft collectives, seed-saver communities and farmers' markets.) Repair is a routine practice in cultures of material scarcity, but within an Australian policy realm that doesn't require product stewardship, or an agreement on responsible product life-cycles, tinkering is a way to extend the life-cycle of products. This is also evident in the rapidly–swelling Visible Mending movement.

Tinkerers' collective behaviour is also informing many nascent policy realms that mediate our consumption and production habits, including Slow Cities, Creative Cities, GovHack and community arts, bike-repair and shed programs. These initiatives have emerged at a time when informal economies and cultural capital are increasingly understood as important for public wellbeing, as public health

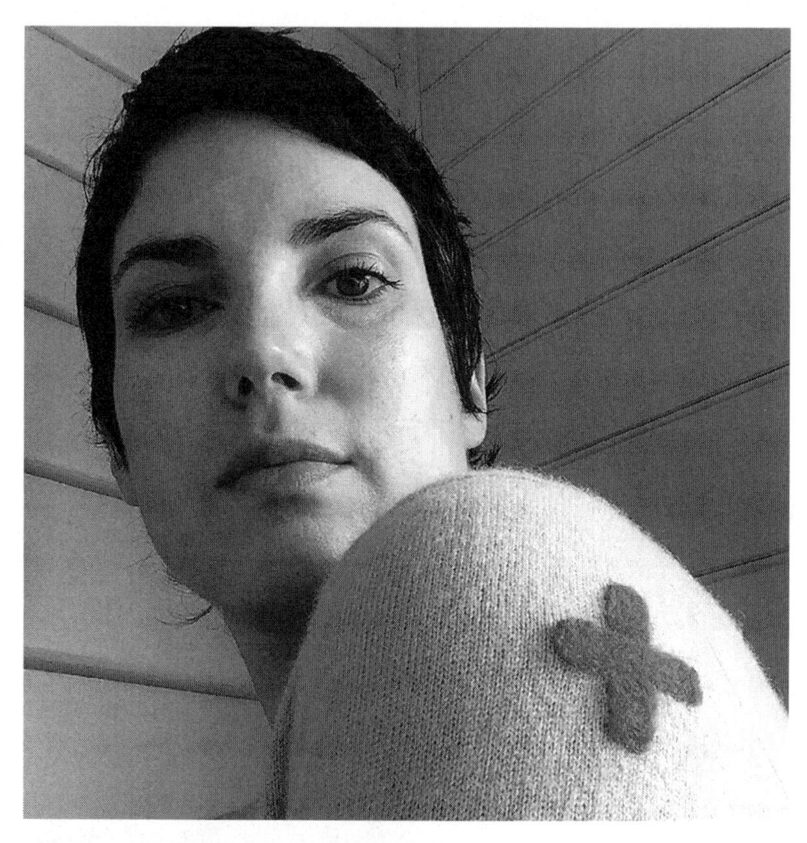

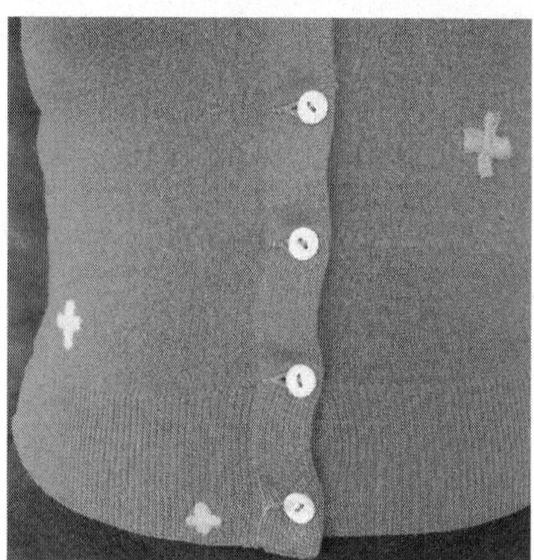

Visible mending projects by Erin Lewis-Fitzgerald, Managing Director at Bright Sparks Australia, a social enterprise that repairs and re-uses discarded appliances.

funding for the development of such projects as FabLabs and Men's Sheds already recognise. The Australian Government's Shed Development Program (AGSDP), which offers funds to more than 1,000 men's sheds across Australia, is 'focused on male health and well-being'.

As studies in urban geography show, practices of material repair often involve struggles between citizens, private industries and public authorities, opening up legal frontiers around issues of duty of care and rights of access. Regulatory attempts to prohibit tinkering effectively deny us the right to own and control many of our consumer products. Many of these – especially our communications technologies – license us to use them only in prescribed ways that can damage the planet, lock us into punitive corporate contracts, devalue our agency and decrease our civil liberties (involving privacy, surveillance and data collection issues). Yet tinkerers I interviewed tended to take heroic pride in claiming ownership of products by disregarding shrink-wrap clauses, warrantee waivers, licensing agreements and other regulatory barriers.

This kind of material–civil disobedience can impact industrial design more broadly. According to a research team headed by British sociologist Elizabeth Shove, the DIY hardware-hacker has propelled industrial design and practice beyond *user-friendly* to *people-centred design*. The industrial designer's role has changed from creating fixed products to 'products which people can adapt and shape to their own purpose'; products that are 'open for others to complete'; and products that 'create a context of experience, rather than just a product'. DIY is also a driving force behind changes to the ways consumer technologies are seen: not as products according to demand or supply, and not products-as-identity, but *products-as-process*, in which the design flow of objects is understood as recursive. Process-oriented-design is a realm that requires 'a more extensive understanding of how materials and practices evolve, circulate and disappear, and a more comprehensive view of things with and through which we view our lives'. Manufacturers' recent interest in the *aftermarket* of products, seen as a growing realm for profitability, has been attributed to what scholars are now calling 'the right to repair movement' (iterated in the Maker's Bill of Rights and the MEND*RS Manifesto).

Stories in this book show how even the most autonomous tinkering develops through the activities of many actors, and also of commons and community resources. In *The Craftsman*, Sennett emphasises the centrality of social connection to good craftsmanship; he supports the medieval idea that craftsmanship requires 'joined skill in community' through 'authority in the flesh'. He also observes that 'all skills, even the most abstract, begin as bodily practices; second, that technical understanding develops through the powers of imagination'. This can transfer to the political imagination.

Yet most of the tinkering in this book occurs in the relatively private domain of home – so the potential for broader cultural impact is difficult to tally. Still, everyday tinkering's impacts are explicit in the stories of the industrial products we engage with, such as the story of the Xbox. In 2001, Microsoft released the Xbox game console to compete with the Sony PlayStation and Nintendo GameCube. As a competitive strategy, Microsoft priced the console at rock-bottom, undercutting competitors and planning for the serious money to be made with the Xbox's exclusive software. To prevent the Xbox from being used with other software, copied games, or alternative operating systems, Microsoft built a *hardware* security system. This proved a red-rag to open source and Linux tinkerers, homebrew developers, game companies and code crackers, who collectively shared information and twice outsmarted Microsoft by tinkering with the untinkerable. 'Because they were all locked out by the same protection, they worked together, either explicitly, or implicitly, by using the results of each other,' wrote Michael Steil, who 'hacks operating systems for a living'. He wrote: 'No Linux hackers ever attacked the PlayStation. When you are fair, people don't fight you.' When engineer Andrew Huang signed up with Wiley to publish *Hacking the Xbox: an Introduction to Reverse Engineering*, Microsoft's lawyers spooked the publisher. Huang self-published nonetheless, and enjoyed instant success. After this, Bill Gates – perhaps looking to value-add his products for the hacker market – reportedly asked: 'How can we engage this [hacking] community?'

So the tinkering mindset can work as a cultural and industrial interloper. And tinkerers' unorthodox engagements with consumer technologies (and thus with power structures) have reportedly inspired the decline and revision of one of capitalism's most destructive material practices – planned obsolescence. According to *Make* editor Mark Frauenfelder, demands generated from tinkering cultures have pressured manufacturers to rethink how they design, regulate and market their products. Tinkerers' habits of material–civil disobedience are key here. Adis Hondo's story especially affirmed what many studies of informal economies have found: innovation is very often the product of unlawful, informal and unregulated activity. The work of economist Colin C Williams is especially valuable in this field, and policy researchers including Charles Leadbeater and Paul Miller have also found that innovation 'often starts in marginal, experimental markets rather than mainstream mass markets' and succeeds where industry can't, in part because 'dedicated amateurs pursue new ideas even when it appears there is no money to be made'. Innovation is also incremental – it's not the heroic light-bulb story that Silicon Valley culture tends to invoke. Throughout industrial history, our most valuable innovation is a cumulative process that tends to come from the ground up, as Chapter 2 discussed.

In Australia, the tinkerer's home isn't a fully private space. Few contemporary Australian homes are. With the proliferation of global communication technologies, our homes are now our workspaces, and places where we do our banking, pay our bills, sign our petitions, do our shopping and trading. They're places we can collectively organise with people we've never met, in shared missions to hack the Xbox, or to demonstrate outside parliament. With social media, our homes have become the public square and the pub; and equally, our habits are increasingly surveilled by spyware embedded in consumer technologies and platforms. Still, Australian homes remain relatively unstructured domains that offer some forms of freedom (free time, free choice, autonomy). In free spaces we can experience tinkering as a way to develop deep knowledge by focusing on small material pursuits. This is the kind of scholarship that isn't possible in bureaucratic institutions, including many education institutions that channel students into our so-called 'knowledge economy'. Pressure to perform, formalised lists of competencies to check off and strictly scheduled and regulated regimes leave little room for the fluidity and freedom of the tinkering method.

Tinkerers' stories reveal tinkering as a way of home-making and meditative connection; as an insular sanctuary from outside impositions of capital and formal institutions. Although the tinkerers I've described actively engage with global capital and its systems (online trade, for example), tinkering is a form of autonomous production not entirely reducible to (or dependent upon) the currents of capital or the fluxes of the labour market. If a component isn't available through online trade, there are always other strategies. Tinkering's resourcefulness makes it ductile and mutable, and most tinkerers I met tended to choose paid work within creative fields or casual structures that allow them to engage with life and time as a tinkerer. The value of this can't be tallied within orthodox approaches to economics, because tinkering isn't simply a productive activity or a creative industry, but an approach to time and life as a whole. Tinkering at home is as much about custom-building objects as about custom-building a meaningful life.

The persistence of tinkering in a post-industrial country like Australia, at a time of cheap and readily-available mass-produced goods, suggests that the impulse to tinker is motivated by a relationship with materiality that can't be reduced to thrift, economy or material need. We know from the work of sociologists and economists (most notably Colin C Williams) that DIY is generally not solely motivated by these concerns. And nor can tinkering be classed simply as a leisure pursuit, or as a triumph of DIY marketing: tinkering projects' discreteness and specificity tend to entail unorthodox, hybrid and sometimes illegal DIY solutions. The projects described in this book often *devalue* commodity forms in formal economic terms, but tinkerers also add layers of personal meaning, currency and memories beyond those that commodity forms can offer. The freedoms these tinkerers seek are not the freedoms associated with free-market and neoliberal values.

One morning, when my waking thoughts panned over years of fieldnotes in which I documented tinkerers' homes, it struck me that most of my research participants had no – or very low – front fences. Rural tinkerers had breezy wire fences to contain livestock, but most of the inner- and outer-urban tinkerers didn't have clearly defined borders between their home- and street-lives. In a culture where front fences are fixtures in the urban landscape, their gardens tended to spill onto the civic sphere. At the time of my doctoral research, I wasn't astute enough to notice this and ask why, but it's tempting now to think of 'tinkerer' as a synecdoche for a wider culture. It may be wishful projection; it may be a long bow to draw, but this fenceless detail now seems meaningful at a time when there's populist momentum to 'build a wall' between nation-states, at a time when 65 million refugees are seeking homes globally. The tinkerers I studied hadn't fortified their homes; their properties instead expressed the values of 'open' (open source; open software; open access; open gardens; open society; sharing economy; transparent government). In *The Biggest Estate on Earth* (2011), which debunks the myth that Australia was an unfarmed wilderness before European occupation, historian Bill Gammage describes why settler-descendant farmers find it difficult to conceive of our landscape – or of agriculture – as a commons and a continuum, in ways First Australians did. 'Fences on the ground,' he asserts, 'create fences in the mind'. A whole lineage of theorists describe how, as Langdon Winner puts it, 'artifacts can contain political properties'. Tasmanian philosopher Jeff Malpas maintains that we rehearse memories, conversations and metaphors in our architectural constructions. Estonian ethnographer Francisco Martínez describes people whose political commitments are expressed through their material domains, and other scholars (as I've discussed) examine the utopian narratives embedded in design, suggesting that these narratives and values are generated from material formations throughout society. The tinkerers who inhabited these fenceless homes didn't feel vulnerable or insecure amid our culture's disruptive work conditions, because their tinkering was a source of security and identity against the impositions of deregulated job markets, competition and labour alienation. These are the very forces that drive the populist sentiment that's displacing people and erecting barriers between them. Some researchers, such as Tony Blair's former policy-adviser Charles Leadbeater, are arguing for 'a new kind of shared home economics, of home-making and building. The route to power to change society starts at home.'

Abandoned road grader on the outskirts of Port Augusta. Photograph by Mark Thomson.

It's an attractive idea, but it's also important not to overstate the revolutionary promise of domestic tinkering, as so many Silicon Valley types have, in flamboyant, heroic, nationalistic and entrepreneurial rhetoric. The tinkering that occurs in Australian homes might be regarded as politically quietist, but it's a certain material provocation, and tinkerers are producers of political and moral as well as human expression. The people in this book reveal tinkering as a way to reinscribe artisanal knowledge into mass culture. For the most part, their practices bypass the formal economy and its values – but then they influence these things as well. Their work relocates value in the *making* of things rather than the commodity-product. By extension, they forge a path to self-sufficiency and agency. In doing these things, they challenge the atomisation of capitalist society.

So I hope this book has shown how tinkering's values of production themselves promote the production of values. Freedom, sanctuary, home-making, risk-taking, scholarship, self-understanding, community-building, democratic self-reliance, artisanal pride, low environmental impact, reciprocity – these values move in recursive ways that circulate through material, conceptual, social, kinaesthetic and temporal forms. The tinkering mindset is continually modified through the body, through practice, through production and product – and also through attendant ideas, technologies and social discourses. Everyday tinkering, then, is the grassroots materialisation of our values. Whether practised in prisons, in warzones, or in peacetime at home, tinkering is a foil to conditions of power. As such, it's an embodiment of hope.

Select References

ABC. *Stateline*. South Australia, presented by Brian Dawe, 4 June 2010.

ABC Radio National. *Artworks*, 30 May 2010.

ABC Radio National. "Tinkering and the Back Yard", *Future Tense*, 19 November 2009, http://www.abc.net.au/radionational/programs/futuretense/tinkering-and-the-backyard/3086078 (2009).

ABC Radio National. "Tinkering with the Future", *Future Tense*, 29 October 2009, http://www.abc.net.au/radionational/programs/futuretense/tinkering-with-the-future/3098980 (2009).

ABC Radio National. "Tinkering and Product Longevity", *Future Tense*, 15 November 2009, http://www.abc.net.au/rn/futuretense/stories/2009/2728886.htm (2009).

Abelon, Dan. "Hacker Culture: The Key to Future Prosperity?", *TechCrunch*, 5 March 2002.

Abrams, Rebecca. *The Playful Self*. Fourth Estate, 1997.

Adamson, Glenn, editor. *The Craft Reader*. Berg, 2010.

Adamson, Glenn. *The Invention of Craft*. Bloomsbury, 2013.

Adamson, Glenn. *Thinking Through Craft*. Oxford: Berg, 2007.

Adamson, Glenn, Giorgio Riello and Sarah Teaseley, editors. *Global Design History*. Routledge, 2011.

Adorno, Theodor. *Aesthetic Theory*. Bloomsbury Academic, 2001.

Allon, Fiona. *Renovation Nation: Our Obsession With Home*. University of New South Wales Press, 2008.

American Honda Motor Co Inc. "Handmade Series DIY Car", available at http://www.slate.com/blogs/future_tense/2014/04/02/honda_april_fools_day_ad_honda_homemade_series_spoof_mocks_the_diy_fad.html (2 April 2004).

Anderson, Chris. *Makers: The New Industrial Revolution*. Random House, 2012.

Appadurai, Arjun. *The Social Life of Things*. Cambridge University Press, 1988.

Arendt, Hannah. *The Human Condition* (second edition). University of Chicago Press, 1998.

Armin, Ash, Angus Cameron and Ray Hudson. "The Alterity of the Social Economy", *Alternative Economic Spaces*, edited by Andrew Leyshon, Roger Lee and Colin C. Williams, 27-54. SAGE, 2003.

Atherton, Cassandra L. *Flashing Eyes and Floating Hair: A Reading of Gwen Harwood's Pseudonymous Poetry*. Australian Scholarly Publishing, 2006.

Atkinson, Paul. *The Ethnographic Imagination: Textual Constructions of Reality*. London: Routledge, 1990.

Atkinson, Paul, Amanda Coffey, Sarah Delamont and John Lofland, editors. *Handbook of Ethnography*. SAGE, 2002.

Back, Les. *The Art of Listening*. Berg, 2007.

Balsamo, Anne. *Designing Culture: The Technological Imagination at Work*. Duke University Press, 2011.

Banks, Marcus. *Visual Methods in Social Research*. SAGE, 2001.
Barry, Andrew. *Political Machines*. The Athlone Press, 2001.
Baum, Scott and Riaz Hassan. "Home Owners, Home Renovation and Residential Mobility", *Journal of Sociology*, vol. 35, no. 1 (1999): 23-41.
Bauman, Zygmunt. *Liquid Modernity*. Polity, 2000.
Bauman, Zygmunt. *Wasted Lives: Modernity and its outcasts*. John Wiley, 2003.
Beadry, Mary C. and Dan Hicks. *The Oxford Handbook of Material Culture Studies*. Oxford University Press, 2010.
Bebergal, Peter. "The Age of Steampunk: Nostalgia Meets the Future, Joined Carefully with Brass Screws", *The Boston Globe*, 26 August 2007.
Beck, Ulrich. *The Brave New World of Work*. SAGE, 2000.
Beck, Ulrich. *Risk Society*. Cambridge University Press, 1992.
Becker, David. "Testing Microsoft and the DMCA", *cnet*, www.cnet.com/au/news/testing-microsoft-and-the-dmca/, 15 April 2003.
Becker, Howard. "Field Methods and Techniques: A Note on Interviewing Tactics", *Interviewing*, edited by N. Fielding. SAGE, 2003: 45-8.

Bennett, Jane. *Vibrant Matter: a political ecology of things*. Duke University Press, 2010.
Bennett, Tony. "The Exhibitionary Complex", *Culture/Power/History: A Reader in Contemporary Social Theory*, edited by Nicholas B. Dirks, Geoff Eley and Sherry B. Ortner. Princeton University Press, 1994.
Bijker, Wiebe. *Of Bicycles, Bakelites and Bulbs: Toward a Theory of Social Change*. Cambridge MA: MIT Press, 1995.
Binfield, Kevin. *Writings of the Luddites*. Baltimore: The Johns Hopkins University Press, 2004.
Binns, Lily and Patrick Buckley. *The Hungry Scientist Handbook: Electric Birthday Cakes, Edible Origami, and Other DIY Projects for Techies, Tinkerers and Foodies*. Collins Living, 2008.
Birtchnell, Thomas and John Urry. "Fabricating Futures and the Movement of Objects", *Mobilities,* vol. 8 no. 3 (2013): 388-405.
Blaxter, Lorraine, Christina Hughes and Malcolm Tight. *How to Research* (Second Edition). Open University Press, 2001.
Bond, Steven, Caitlin DeSilvey, and James R. Ryan. *Visible Mending: Everyday Repairs in the South-West*. Uniform Books, 2013.
Bongiorno, Frank. "Two Radical Legends: Russel Ward, Humphrey McQueen and The New Left Challenge in Australian Historiography", *Russel Ward: Reflections on a Legend*, edited by Bongiorno, Frank and Roberts, David Andrew, *Journal of Australian Colonial History*, v.10, no.2 (2008): 201-22.
Bourdieu, Pierre (trans. R Nice). *Distinction: A Social Critique of the Judgement of Taste*, translated by R Nice. Harvard University Press, 1984.
Bourdieu, Pierre. *The Logic of Practice*. Stanford University Press, 1990.
Boym, Svetlana. *The Future of Nostalgia*. Basic Books, 2001.
Bracey, Catherine. "Why Good Hackers Make Good Citizens", TED TALKS September 2013.

Select References

Bratich, Jack Z. and Heidi M. Brush. "Fabricating Activism: Craft-Work, Popular Culture, Gender", *Utopian Studies,* vol. 22 no. 2 (2011): 233-60.

Brennan, Teresa. *The Transmission of Affect.* Cornell University Press, 2004.

Brett, Judith. *Quarterly Essay 19: Relaxed & Comfortable: The Liberal Party's Australia.* Black Inc., 2005.

Brown, Stuart. *Play: How it Shapes the Brain, Opens the Imagination, and Invigorates the Soul.* Scribe, 2009.

Brownlee, John. "Meet Mr Steampunk: Jake von Slatt", *Wired,* 29 July 2007.

Burgess, Jean. "Defining Vernacular Creativity", *Creativity Machine Weblog.* Available at: http://www.creativitymachine.net (2006).

Burke, Mary M. *Tinkers: Synge and the Cultural History of the Irish Traveller.* Oxford University Press, 2009.

Buscher, Monika and Leon Cruikshank. "Designing Cultures: Post-disciplinary Practices", paper delivered at the 8th European Academy of Design Conference, Aberdeen, Scotland, 2009.

Callon, Michel, editor. *The Laws of the Markets.* Blackwell, 1998.

Campbell, Colin. "The Craft Consumer: Culture, Craft and Consumption in a Postmodern Society", *Journal of Consumer Culture,* vol. 5 no. 1 (2005): 23-42.

Carey, John. *What Good Are The Arts?.* Oxford University Press, 2006.

Carson, Kevin. *Desktop Manufacturing: A Homebrew Industrial Revolution.* Centre for a Stateless Society, 2009.

Carter, David. "Going, Going Gone? Britishness and Englishness in Contemporary Australian Culture", *Overland,* vol. 169 (Summer 2002): 81-6.

Carter, Nanette. "Blueprint to patterncraft: DIY furniture patterns and packs in post-war Australia", *Design Activism and Social Change,* Conference of the Design History Society, 2011:1-11.

Catastrophone Orchestra and Arts Collective. "What Then, Is Steampunk?", *Steampunk Magazine.* 1 September, 2007.

Caterson, Simon. *Hoax Nation.* Arcade Publications, 2010.

Chabaud-Rychter, Danielle. "The Configuration of Domestic Practices in the Designing of Household Appliances", *The Gender-Technology Relation: Contemporary Theory and Research,* edited by Gill and Grint. Taylor & Francis, 1995: 95-111:

Challa, Janaki. "Why Being 'Gypped' Hurts The Roma More Than It Hurts You", *Codeswitch: Frontiers of Race, Culture and Ethnicity,* National Public Radio, 30 December, 2013.

Chaplin, Lan Nguyen and Deborah Roedder John. "Growing Up in a Material World: Age Differences in Materialism in Children and Adolescents", *Journal of Consumer Research,* vol. 34 no. 4 (2007): 480-93.

Chaplin, Lan Nguyen and Deborah Roedder John. "Interpersonal Influences on Adolescent Materialism: A New Look at the Role of Parents and Peers", *Journal of Consumer Psychology,* vol. 20 no. 2 (2010): 176-84.

Chapman, Jonathan. *Emotionally Durable Design.* Earthscan, 2005.

Chesney, Kellow. *The Victorian Underworld*. Penguin, 1970.
Cieraad, Irene. *At Home: An Anthropology of Domestic Space*. Syracuse University Press, 1999.
Connor, Steven. *Paraphernalia: The Curious Life of Magical Things*. Profile Books, 2011.
Costello, Patrick. *Action Research*. A & C Black, 2003.
Crawford, Matthew B. "Manual Competence", TED-X, 9 May 2011.
Crawford, Matthew B. *Shop Class as Soulcraft: An Inquiry into the Value of Work*. Penguin, 2009.
Critchley, Christine and Gilding, Michael. "Technology and Trust: Public Perceptions of Technological Change in Australia", *Australian Journal of Emerging Technologies and Society*, vol. 1 no. 1 (2003): 52-69.
Csikszentmihalyi, Mihaly. *Flow: The Psychology of Optimal Experience*. New York: Harper Perennial Modern Classics, 2008.
Cunningham, Stuart. *Hidden Innovation: Policy, Industry and the Creative Sector*. University of Queensland Press, 2013.
Daniels, Steve, editor. *Makeshift*, no. 1, Fall (2011).

Dant, Tim. *Material Culture in the Social World*. Open University Press, 1999.
Darian-Smith, Kate. *On The Home Front: Melbourne in Wartime, 1939-1945* (Second Edition). MUP, 2009.
Dawkins, Nicole. "Do-It-Yourself: The Precarious Work and Postfeminist Politics of Handmaking (in) Detroit", *Utopian Studies*, vol. 22 no. 2 (2011): 261-84.
De Certeau, Michel. *The Practice of Everyday Life*, translated by Steven F Rendall. University of California Press, 1984.
Deci, Edward and Richard Ryan. "Hedonia, Eudaimonia and Wellbeing: An Introduction", *Journal of Happiness Studies,* vol. 9 no. 1 (January 2008): 1-11.
Delamont, Sara and Paul Atkinson. "Doctoring Uncertainty: Mastering Craft Knowledge", *Social Studies of Science,* vol. 31 no. 1 (2001): 87-107.
Delfanti, Alessandro. *Biohackers: The Politics of Open Science*. Pluto Press, 2013.
Dessaix, Robert. "Loitering with Intent: Reflections on the Demise of the Dilettante", *Seams of Light: Best Antipodean Essays*, edited by Morag Fraser. Allen & Unwin, 1998.
Dezeuze, Anna. "Assemblage, Bricolage and the Practice of Everyday Life", *Art Journal,* vol. 67 no. 1 (2008): 31-7.
Diamond, Jared. *Guns, Germs, and Steel: The Fates of Human Societies.* Norton (1997).
Diani, Mario. "The Concept of Social Movements", *Sociological Review*, vol. 40 no. 1 (1992). 1-25.
Dickens, Charles. *A Tale of Two Cities*. Penguin Classics, 2003.
Dielter, Michael. "Consumption", *The Oxford Handbook of Material Culture Studies*, edited by Dan Hicks and Mary Beaudry. Oxford University Press, 2010.
Dixon, Stephen. "Beyond the Seas", *craft + design enquiry,* no. 1 (2009): 1-30.
Doctorow, Cory. "Love the Machine, Hate the Factory", *MAKE* 17 (undated): 14.
Doctorow, Cory. *Makers*. Tom Doherty Associates, 2009.
Doherty, Dale. Foreword to *The Art of Tinkering: Meet 150+ Makers Working at the Intersection of Art and Science*, edited by Mile Petrich, and Karen Wilkinson, no. 6.

Weldon Owen, 2014.
Dormer, Peter. *The Art of the Maker*. Thames and Hudson, 2004.
Dovey, Kim. "Dreams on Display: Suburban Ideology in the Model Home", *Beasts of Suburbia: Reinterpreting Cultures in Australian Suburbs*, edited by Chris Healey, Sara Ferber and Chris McAuliffe. Melbourne University Press, 1994:127-47.
Doyle, Brown. "A Tinkering Kind Of Guy", *American Scholar*, 15 May 2015.
Dreher, Rod. *Crunchy Cons*. Crown Forum, 2006.
Duguid, Paul. "The Art of Knowing: Social and Tacit Dimensions of Knowledge and the Limits of the Community of Practice", *The Information Society: An International Journal*, vol. 21 no. 2 (2005): 109-18.
Durbin, Jonathan. "The Artful Wit of Steampunk: A Retro Design Movement That's All about the Future", *Papermag*, 2 May 2008.
Edgerton, David. *The Shock of the Old: Technology and Global History Since 1900*. Profile Books, 2006.
Eglash, Ron. "Technology as Material Culture", *Handbook of Material Culture*. SAGE, 2006.
Eilam, Eldad. *Reversing: Secrets of Reverse Engineering*. Wiley, 2005.
Entwistle, Joanne and Don Slater. "Reassembling the Cultural: Fashion Models, Brands and the Meaning of 'Culture' after ANT", *Journal of Cultural Economy*, vol. 7 no. 2 (2014): 161-77.
Fairclough, Norman. *Analysing Discourse: Textual Analysis for Social Research*. Routledge, 2003.
Faricy, Michele and Christopher Hoyman. "It Takes a Village: A Test of the Creative Class, Social Capital and Human Capital Theories", *Urban Affairs Review* vol. 44 no. 3 (2009).
Fearn-Wannan, Bill. *Australian Folklore: A Dictionary of Lore, Legends and Popular Allusions*. Lansdowne, 1970.
Fixers Collective, available at http://www.fixerscollective.org/.
Fodor, Jerry. *Hume Variations*. Oxford University Press, 2003.
Foege, Alec. *The Tinkerers: The Amateurs, DIYers, and Inventors Who Make America Great*. Basic Books, 2013.
Fox, Kathryn Joan. "Real Punks and Pretenders: The Social Organisation of a Counter-Culture", *Journal of Contemporary Ethnography*, vol. 16 no. 3 (1987): 344-70.
Fox, Nichols. *Against the Machine: The Hidden Luddite Tradition in Literature, Art and Individual Lives*. Island Press, 2002.
Frank, Thomas. *One Market under God: Extreme Capitalism, Market Populism and the End of Economic Democracy*. Doubleday, 2000.
Franz, Kathleen. *Tinkering: Consumers Reinvent the Early Automobile*. Philadelphia: University of Pennsylvania Press, 2005.
Frauenfelder, Mark. *Made By Hand: Searching for Meaning in a Throwaway World*. Portfolio, 2009.
Friedman Jonathan, editor. *Consumption and Identity*. Taylor & Frances, 2005. Taylor & Frances e-library.

Gabrielson, Curt. "Build this Model Arm and Explore Human Tinkering", *Make* 22 April 2014.

Galbraith, Kate. *The Great Texas Wind Rush: How George Bush, Ann Richards, and a Bunch of Tinkerers Helped the Oil and Gas State Win the Race to Wind Power.* University of Texas Press, 2013.

Gammage, Bill. *The Biggest Estate on Earth: How Aborigines made Australia.* Allen & Unwin, 2011.

Garber, Marjorie. *Academic Interests.* Princeton University Press, 2003.

Gauntlett, David. *Making Is Connecting: The Social Meaning of Creativity, From DIY and Knitting to YouTube and Web 2.0.* Polity Press, 2011.

Gelder, Ken. *Subcultures: Cultural History and Social Practices.* Routledge, 2007.

Gell, Alfred. *Art and Agency.* Clarendon Press, 1998.

Gell, Alfred. "The Enchantment of Technology and the Technology of Enchantment", *Handbook of Material Culture.* SAGE, 2006.

Gibson, Katherine and Julie Grahame. *The End of Capitalism (As We Knew It): A Feminist Critique of Political Economy.* University of Minnesota Press, 2006.

Gibson, Mary Jo and Ari Houser. "Valuing the Invaluable: A New Look at the Economic Value of Family Caregiving", AARP Public Policy Institute (2007).

Gibson, William and Bruce Sterling. *The Difference Engine.* Bantam Books, 1991.

Glaser, Barney. "Doing Grounded Theory: Issues and Discussions", *Using Grounded Theory to Interpret Interviews,* Linda Jo Calloway, Constance A. Knapp, available at http://csis.pace.edu/~knapp/AIS95.htm. Sociology Press, 1998.

Gogol, Nikolai. *Dead Souls* (1842).

Google Ngram Viewer, search term "tinker", "tinkering" and "tinkerer" between 1800-2000. Available at http://books.google.com/ngrams/graph?content=tinkering&year_start=1800&year_end=2000&corpus=0&smoothing=3.

Graeber, David. "On the Phenomenon of Bullshit Jobs", *Strike Magazine,* 17 August 2013.

Graeber, David. *Possibilities: Essays on Hierarchy, Rebellion and Desire.* A K Press, 2007.

Graeber, David. *Toward an Anthropological Theory of Value: The False Coin of Our Own Dreams.* Palgrave Macmillan, 2001.

Graham, Stephen and Nigel Thrift. "Out of Order: Understanding Repair and Maintenance", *Theory, Culture Society,* vol. 24 no. 1 (2007): 1-25.

Gray, Stephen. "Going Native: Disguise, Forgery, Imagination and the 'European Aboriginal'", *Overland* no. 170 (2003): 34-42.

Greenland, Hall. "A Green by Any Other Name", *The Greens Magazine,* 13 August 2014, http://greens.org.au/magazine/national/green-by-any-other-name.

Greenop, Matt. "Sony's 'Unhackable' PlayStation 3 hacked", *The Independent,* 26 January 2010, http://www.independent.co.uk/life-style/gadgets-and-tech/news/sonys-unhackable-playstation-3-hacked-1879215.html.

Greer, Betsy. "Craftivist History", *Extra/Ordinary: Craft and Contemporary Art,* edited by Maria Elena Buszek. Duke University Press, 2011.

Gregg, Melissa. "Feeling Ordinary: Blogging as Conversational Scholarship", *Continuum: Journal of Media & Cultural Studies,* vol. 20 no. 2 (2006): 147-60.

Gregg, Melissa. *Work's Intimacy*. Polity Press, 2011.
Gregg, Melissa and Gregory Seigworth, editors. *The Affect Theory Reader*. Polity Press, 2010.
Gregson, Nicky, Alan Metcalfe and Louise Crewe. "Practices of Object Maintenance and Repair: How Consumers Attend to Objects Within the Home", *Journal of Consumer Culture*, vol. 9 no. 2 (2009): 248-72.
Gross, Cori. "A History of Steampunk: Part IV - A Genre Comes of Age", *Voyages Extraordinaire: Scientific Romances in a Bygone Era*, 19 August, 2008. Accessed 16 April 2009, http://voyagesextraordinaires.blogspot.com/2008/08/history-of-steampunk-part-iv-genre.html.
Grosz, Elizabeth. *Chaos, Territory, Art: Deleuze and the Framing of the Earth*. Columbia University Press, 2008.
Gruppetta, Maree. "Snowball Recruiting: Capitalising on the Theoretical 'Six Degrees of Separation'", paper delivered at AARE Conference, 2005.
Guizzo, Erico. "The Steampunk Contraptors," *IEEE Spectrum*, vol. 10 no. 8 (October 2008): 43-9.
Hackney, Fiona. "Quiet Activism and the New Amateur: The Power of Home and Hobby Crafts", *Design and Culture*, vol. 5 no. 2 (2013):169-94.
Hagel, John, John Seely-Brown and Duleesha Koolasuriya. *A Movement in the Making*. Deloitte University Press, 2014.
Hall, T. *Where Have All the Gardens Gone? An Investigation into the Disappearance of Backyards in the Newer Australian Suburb*. Brisbane: Urban Research Program, Griffith University, 2007.
Hamelink, Cees J. "Prometheus in Cyberspace', in *The Ethics of Cyberspace*, Sage, London (2000): 3.
Hamilton, Clive and Elizabeth Mail. 'Downshifting in Australia: A Sea Change in the Pursuit of Happiness", *Discussion Paper 50*. The Australia Institute, 2003.
Hamilton, Clive and Richard Denniss. *Affluenza: When Too Much is Never Enough*. Allen & Unwin, 2005.
Hampâté Bâ, Amadou. "African Art: Where the Hand Has Ears", UNESCO *Courier* (February 1976).
Hårdm, Mikael and Ruth Oldenziel. *Consumers, Tinkerers, Rebels: The People Who Shaped Europe*. Palgrave Macmillan, 2013.
Harkinson, Josh. "The Unbearable Whiteness of Silicon Valley", *Mother Jones* (July 2015): 24-8.
Harrison, Charles, Paul Wood and Jason Gaiger, editors. *Art in Theory, 1815-1900: An Anthology of Changing Ideas*. Carlton: Blackwell Publishing, 1998.
Harvey, Karen, editor. *History and Material Culture*. New York: Routledge, 2010.
Hatch, Mark. *The Maker Movement Manifesto: Rules for Innovation in the New World of Crafters, Hackers, and Tinkerers*. McGraw Hill, 2014.
Hawkins, G. *The Ethics of Waste*. Oxford: Roman and Littlefield Publishers Inc., 2005.
Hawkin, Paul, Amory B. Lovens, and Hunter L. Lovens. *Natural Capitalism: The Next Industrial Revolution*. London: Earthscan Publications, 1999.

Hayward, Philip, editor. *Culture, Technology and Creativity in the Late Twentieth Century*. John Libbey and Company, 1990.

Heath, Joseph and Andrew Potter. *The Rebel Sell: Why the Culture Can't Be Jammed*. Toronto: HarperCollins, 2004.

Hebdige, Dick. *Subculture: The Meaning of Style*. Routledge, 1979.

Hell, Kyshah. "Clockworks and Carbon: The Fantastical Escapades of the Steampunk Aesthetic", *Morbid Outlook Magazine*, July 2009. Accessed 21 July 2009, http://morbidoutlook.com/fashion/articles/2009_07_steampunk.html.

Hemming, Steve. "Objects and Specimens: Conservative Politics and the SA Museum's Aboriginal Cultures Gallery", *Overland*, no. 171 (Winter 2003): 64-9.

Hesmondhalgh, David. "User-generated Content, Free Labour and the Cultural Industries", *Ephemera: Theory & Politics in Organisation* no. 10 (2010): 267-84.

Hesmondhalgh, David and Sarah Baker. *Creative Labour: Media Work in Three Cultural Industries*. Routledge, 2011.

Hickman, Pekka. *The Hacker Ethic: A Radical Approach to the Philosophy of Business*. Random House Trade, 2001.

Hicks, Dan. "The Material Cultural Turn", *The Oxford Handbook of Material Cultural Studies*. OUP (2010): 25-98.

Highmore, Ben. "Bitter After Taste: Affect, Food and Social Aesthetics", *The Affect Theory Reader*, edited by Melissa Gregg and Gregory Seigworth. Polity Press, 2010.

Highmore, Ben. *Everyday Life and Cultural Theory: An Introduction*. Taylor & Francis, 2002.

Hirst, John. *Looking for Australia: Historical Essays*. Black Inc., 2010.

Hobbs, AC. "The Lock Question", Letter to the Editor, *The Times*. 15 July 1853. Cited in Glenn Adamson, *The Invention of Craft*. Bloomsbury, 2013.

Hodgkinson, Tom. *How to Be Idle*. Penguin, 2005.

Hoyman, Michele and Christopher Faricy. "It Takes a Village: A Test of the Creative Class, Social Capital and Human Capital Theories", *Urban Affairs Review*, January, vol. 44 no. 3 (2009): 311-333.

Huggan, Graham. *Postcolonial Exotic: Marketing the Margins*. Routledge, 2001.

Hunter, Dan, Ramon Lobato, Megan Richardson and Julian Thomas, editors. *Amateur Media: Social, Cultural and Legal Perspectives*. London: Routledge, 2013.

Hut, Rolf, speaking on TED Talks, 23 January 2012.

Ingold, Tim. *Making: Anthropology, Archeology, Art and Architecture*. Routledge, 2013.

Ingram, Jack, Elizabeth Shove, Matthew Watson and Martin Hand, editors. *The Design of Everyday Life*. Berg, 2007.

Instructables. *Make, How, Do and DIY*, available at: http://www.instructables.com (2007).

Ito, Mimi. "Amateur Cultural Production and Peer-to-Peer Learning", paper delivered at the American Educational Research Association, 2008.

Ito, Mimi. *Imagining the Future of Amateur Cultural Production*, blog post 28 February, available at: http://www.itofisher.com (2006).

Jackson, Michael. *At Home In The World.* Duke University Press, 1995.
Jackson, Simon. "Sacred Objects – Australian Design and National Celebrations", *Journal of Design History,* vol. 19 no. 3 (2006): 249-55.
Jacob, Francois. "Evolution and Tinkering", *Science,* vol. 196 no. 4295 (1977): 1161-66.
Jalopy, Mister. "A Maker's Bill of Rights to Accessible, Extensible, and Repairable Hardware", *Make,* no. 4 (2005).
James, Alison, Jenny Hockey and Andrew Dawson, eds. *After Writing Culture: Epistemology and Practice in Contemporary Anthropology.* Taylor and Francis, 2003.
Jencks, Charles and Nathan Silver. *Adhocism: The Case for Improvisation.* New York: Anchor Books, 1972.
Jenkins, Henry. *Fans, Bloggers and Gamers: Exploring Participatory Culture.* New York: New York University Press, 2006.
Jha [no other name provided]. "The Intersection of Race and Steampunk: Colonialism and Its After-effects and Other Stories, from a Steampunk of Colour's Perspective" (blog post), *Racialicious,* 24 June (2009).
Johnson, Steven M. *What the World Needs Now: A Resource Book for Daydreamers, Frustrated Inventors, Cranks, Efficiency Experts, Utopians, Gadgeteers, Tinkerers and Just about Everybody Else.* Patent Depending Press, 2012.
Johnson, Steven. *Emergence: The Connected Lives of Ants, Brains, Cities and Software.* New York: Scribner, 2001.
Johnson, Steven. *Where Good Ideas Come From: The Natural History of Innovation.* New York: Riverhead Books, 2010.
Johnston, Fanny. "Meet the Mothers of Invention", *The Guardian Online,* 1 February 2008, accessed 10 October 2008, available at http://www.guardian.co.uk.
Jones, Caroline. *The Machine in the Studio: Constructing the Postwar American Artist.* University of Chicago Press, 1996.
Jones, Steven E. *Against Technology: From the Luddites to Neo-Luddism.* London: Routledge, 2006.
Josephides, Lisette. "Representing the Anthropologist's Predicament", *After Writing Culture: Epistemology and Praxis in Contemporary Anthropology,* edited by Allison James, Jenny Hockey and Andrew Dawson. Routledge, 1997.
Jungnickel, Kat. "Making WiFi: A Sociological Study of Backyard Technologists in Suburban Australia", PhD diss., Studio INCITE, Goldsmiths College, University of London, 2009.
Kalish, Jon. "DIY 'Hackers' Tinker Everyday Things into Treasure", National Public Radio, 21 November 2010.
Kastner, Jeffrey. "National Insecurity", *Cabinet Magazine,* no. 22 (2001). http://cabinetmagazine.org/issues/22/kastner.php.
Keen, A. *The Cult of the Amateur: How Today's Internet is Killing Our Culture and Assaulting our Economy.* London: Nicholas Brealey Publishing, 2007.
Kelly, Julie. "The Archaeology of Assemblage", *Art Journal,* vol. 67 no. 1 (2008): 24-30.
Kelly, Kevin. *What Technology Wants.* New York: Viking, 2010.

Kelty, Christopher M. *Two Bits: The Cultural Significance of Free Software*. Duke University Press, 2008.

Kittler, Friederich A. *Gramophone, Film, Typewriter*. Stanford: Stanford University Press, 1999.

Kline, Ronald J. *Consumers in the Country: Technology and Social Change in Rural America*. John Hopkins University Press, 2000.

Knepper, Wendy. "Colonization, Creolization, and Globalization: The Art and Ruses of Bricolage", *Small Axe*, Indiana University Press, 21 October, vol. 10 no. 3 (2006): 70-86.

König, Anna. "A Stitch in Time," *Culture Unbound,* no. 5 (2013): 569-85.

Korhonen, Kuisma, editor. *Tropes for the Past: Hayden White and the History/Literature Debate*. Editions Rodopi, 2006.

Krause, Elliot. *Death of the Guilds*. Yale University Press, 1996.

Krimsky, Sheldon. 'Introduction', *Edging Towards BioUtopia,* Richard Hindmarsh. University of Western Australia Press, 2008.

Krimsky, Sheldon. "Introduction", *Edging Towards BioUtopia,* Richard Hindmarsh. University of Western Australia Press, 2008.

Kuchera, Ben. "Is Bill Gates Learning from the Mod Community?", *Ars Technica*, 23 June 2006.

La Ferla, Ruth. "Steampunk Moves Between 2 Worlds", *New York Times*, 8 May 2008.

Lahart, Justin. "Tinkering Makes a Comeback amid Crisis", *The Wall Street Journal*, 13 November, 2009. A1.

Lash, Scott. "Risk Culture", *The Risk Society and Beyond: Critical Issues for Social Theory,* edited by Barbara Adam, Ulrich Beck and Joost Van Loon. SAGE, 2000.

Lasky, Julia. "Bushpunk and the Future of Africa", *Change Observer,* 13 January 2011, http://changeobserver.designobserver.com/feature/bushpunk-and-the-future-of-africa/24148/.

Latour, Bruno. "How to Talk About the Body? The Normative Dimension of Science Studies", *Body & Society* (2004): 205-29.

Latour, Bruno. *An Inquiry into Modes of Existence: An Anthropology of the Moderns*. Harvard University Press, 2013.

Latour, Bruno. *Reassembling the Social: An Introduction to Actor-Network-Theory*. Oxford University Press, 2005.

Latour, Bruno. "The Recall of Modernity: Anthropological Approaches", *Cultural Studies Review,* vol. 13 no. 1 (2007).

Laubichler, Manfred. "Tinkering: A Conceptual and Historical Evaluation", paper delivered at the Novartis Found Symposium. 284: 20-9; discussion 29-34, 110-5 (2007).

Law, John. *After Method: Mess In Social Science Research*. New York: Routledge, 2004.

Leach, Dan. "'Our Stars and Stripes", *Overland,* no. 171 (Winter 2003): 107-8.

Leadbeater, Charles. "Nobody is Home", *Aeon*. 30 November 2016. https://aeon.co/essays/why-theres-no-place-like-home-for-anyone-any-more.

Leadbeater, Charles and Paul Miller. *The Pro-Am Revolution: How Enthusiasts are*

Changing Our Economy and Society. London: Davos, 2004.
Lefebvre, Henri. "Work and Leisure in Everyday Life", *The Everyday Life Reader*, edited by Ben Highmore. Routledge, 2002.
Legge, Kate. "Secret Men's Business", *The Weekend Australian Magazine*, 25-26 June 2011: 11-4.
Lemonnier, Pierre. *Technological Choices: Transformation in Material Cultures since the Neolithic*. London: Routledge, 1993.
Lévi-Strauss, Claude. *The Savage Mind: Nature of Human Society*. Chicago: University of Chicago Press, 1962.
Lloyd, Brian. "Ern Malley and His Rival", *Australian Literary Studies*, vol. 20 no. 1 (2001): 20.
Logan, Jim. *Everyday Art: Australian Folk Art*. Canberra: National Gallery of Australia, 1998.
Lohrey, Amanda. "The Project of Self in Late Capitalism", *Overland*, no. 164 (Spring 2001): 4-14.
Lovink, Geert. "Christopher Kelty on the Culture of Free Software", *p2p Foundation*, 23 August 2008, http://p2pfoundation.net/Christopher_Kelty_on_the_Culture_of_Free_Software.
Luckman, Susan. *Locating Cultural Work: The Politics and Poetics of Rural, Regional and Remote Creativity*. Macmillan, 2012.
Malkogeorgou, Titika. "Folding, Stitching, Turning: Putting Conservation into Perspective", *Journal of Material Culture*, no. 16 (2011): 441-55.
Mallett, Shelley. "Understanding Home: A Critical Review of the Literature", *The Sociological Review* (2004): 62-89.
Manjoo, Farhad. "The Death of Planned Obsolescence", *Slate* magazine (11 August 2008).
Marcus, George E. "From Rapport under Erasure to Theaters of Complicit Reflexivity", *Qualitative Inquiry*, no. 7 (2001): 519-24.
Martínez, Francisco. "The Ordinary Effects of Repair", *Eurozine*, 16 March 2017.
Martinez, Sylvia Libow, and Gary Stager. *Making, Tinkering and Engineering in the Classroom*. Constructing Modern Knowledge Press, 2013.
Marwick, Alice and Antony Funnell. "Silicon Valley and the Myth of Meritocracy,' *ABC Radio National*, March 18, 2014.
Marx, Karl. "On Alienation (1844)", *Art in Theory, 1815-1900: An Anthology of Changing Ideas*, edited by P.W. Charles Harrison and Jason Gaiger, Carlton: Blackwell, 1998.
Maurer, Bill. "In the Matter of Marxism", *Handbook of Material Culture*, edited by Christopher Tilley, Webb Keane, Susanne Kuechler-Fogden, Mike Rowlands and Patricia Spyer, 13-28. SAGE, 2006.
Mauss, Marcel. *A General Theory of Magic*. Psychology Press, 2001.
McAdams, Dan. *The Redemptive Self*. Oxford University Press, 2005.
McCarthy, AJ. "Honda Salvages April Fools' Day While Skewering DIY Ridiculousness", *Slate* magazine (2 April 2014).
McCray, W Patrick. "It's not all lightbulbs", *Aeon*, 12 October 2016, https://aeon.co/essays/most-of-the-time-innovators-don-t-move-fast-and-break-things.

McCullough, Malcolm. *Abstracting Craft: The Practiced Digital Hand.* Cambridge: The MIT Press, 1996.

McKernan, Michael. *All in! Australia during the Second World War.* Melbourne: Nelson, 1983.

McLeod Kevin. *Kevin McCloud's Man Made Home*, episodes 3 and 4, available at http://www.channel4.com/programmes/kevin-mcclouds-man-made-home.

McLuhan, Marshall. *Understanding Media: The Extensions of Man.* Ginko Press. 2013.

Metcalfe, Andrew and Ann Game. "Creative Practice: The Time of Grace", *Time and Society*, no. 19 (2010): 165-79.

Miller, Daniel. *The Comfort of Things.* Cambridge: Polity Press, 2008.

Miller, Daniel, editor. *Materiality.* Durham: Duke University Press, 2005.

Miller, Daniel. *Stuff.* Cambridge: Polity, 2010.

Mills, Stephanie, editor. *Turning Away from Technology: A New Vision for the 21st Century.* San Francisco: Sierra Club Books 1997.

Minahan, Stella and Julie Wolfram Cox. "Stitch'nBitch: Cyberfeminism, a Third Place, and the New Materiality", *Journal of Material Culture* no. 12 (2007).

Minter, Adam, *Junkyard Planet.* Bloomsbury (2015).

Molloy, Bruce. *Before the Interval: Australian Mythology and Feature Films, 1930-1960.* University of Queensland Press, 1990.

Monahan, Torin and Jill A. Fisher. "Benefits of 'Observer Effects': Lessons from the Field", *Qualitative Research*, vol. 10 no. 3 (2010): 357-76.

Morozov, Evgeny. "Making It", *The New Yorker*, 13 January 2014.

Morozov, Evgeny. "The Meme Hustler: Tim O'Reilly's Crazy Talk", *The Baffler,* no. 22 (2013).

Morphy, Howard. "Art as Action, Art as Evidence", *The Oxford Handbook of Material Cultural Studies*, edited by Dan Hicks and Mary Beaudry. Oxford University Press, 2010: 265-90.

Moyal, A. "Invention and Innovation in Australia: The Historian's Lens", *Prometheus*, vol. 5 no. 1 (1987): 93-110.

Murphy Paul, Annie. "In Praise of Tinkering: How the Decline in Technical Know-how is Making Us Think Less", *Time*, 19 October 2011.

Myers, Fred. "'Primitivism', Anthropology, and the Category of 'Primitive Art'", *Handbook of Material Culture,* edited by Tilley et al. (2010): 267-84.

Myers, Fred. "Some Properties of Art and Culture: Ontologies of the Image and Economics of Exchange", *Materiality*, edited by Daniel Miller. Durham: Duke University Press, 2005: 88-117.

Nakamura, Randy. "Steampunk'd, Or Humbug by Design", *Design Observer*, 22 July 2008, accessed 22 June 2009, http://designobserver.com/feature/steampunkd-or-humbug-by-design/7057.

National Museum of Australia. "The Saw Doctor's Wagon", Education Services Australia Limited and the National Museum of Australia, http://nma.gov.au/collections-search/display?app=tlf&irn=61293 (2009).

Nelson, Camilla. "The Invention of Creativity: The Emergence of a Discourse", *Cultural*

Studies Review, vol. 16 no .2 (2010).

Nevins, Jess. "Introduction: The 19th Century Roots of Steampunk", *Steampunk,* edited by Ann Vandermeer, and Jeff Vandermeer. Tachyon, 2008.

Nippert-Eng, C. E. *Home and Work: Negotiating Boundaries through Everyday Life.* London: University of Chicago Press, 1996.

Nutch, Frank. "Gadgets, Gizmos, and Instruments: Science for the Tinkering", *Science, Technology, & Human Values*, vol. 21 no. 2 (1996): 214-28.

O'Connor, E. "Embodied Knowledge: The Experience of Meaning and the Struggle towards Proficiency in Glassblowing", *Ethnography*, vol. 6 no. 2 (2005): 183-204.

Okely, Judith. "Some Political Consequences of Theories of Gypsy Ethnicity", *After Writing Culture: Epistemology and Practice in Contemporary Anthropology*, edited by Allison James, Alison, Jenny Hockey and Andrew Dawson. Taylor and Francis, 2003.

Oldfield, Molly and John Mitchinson. "QI: How Knitting Was Used as Code in WW2", *The Telegraph*, 18 February 2014.

Oliver, Paul G. and Gill Green. "Adopting New Technologies: Self-sufficiency and the DIY Artist", *Games Computing and Creative Technologies: Journal Articles.* Paper 6 (2009), http://digitalcommons.bolton.ac.uk/gcct_journalspr/6.

Onion, Rebecca. "Reclaiming the Machine: An Introductory Look at Steampunk in Everyday Practice", *Journal of Neo-Victorian Studies*, vol. 1 no. 1 (Autumn 2008).

Outka, Elizabeth. *Consuming Traditions: Modernity, Modernism and the Commodified Authentic.* Oxford University Press, 2009.

Pacaud, Jean-François Quilici. "Dominant Representations and Technical Choices: A Method of Analysis with Examples from Aeronautics", *Technological Choices: Transformations in Material Culture since the Neolithic*, edited by Pierre Lemonnier. Routledge, 2002: 399-412.

Pagliassotti, Dru. "Does Steampunk have Politics?", *The Mark of Ashen Wings*, 11 February 2009, http://ashenwings.com/marks/2009/02/11/does-steampunk-have-politics/.

Painter, Colin, editor. *Contemporary Art and the Home.* Berg, 2002.

Pang, Alex. "Reflections on Tinkering", *Relevant History* (blog) available at http://askpang.typepad.com/relevant_history/2008/10/reflections-on.html, 5 October (2008B).

Pang, Alex. "Tinkering," *Relevant History* (blog), http://askpang.typepad.com/relevant_history/2008/10/tinkering.html, 28 October (2008A), reporting on the conference, "Tinkering as a Mode of Knowledge: Production in the Digital Age".

Park, Miles. "Defying Obsolescence." *Longer Lasting Products: Alternatives to the Throwaway Society*, edited by T. Cooper. (Kingston University, 2009). 2010: 77-95.

Park, Miles. "Product Life: Designing for Longer Lifespans", PhD diss., Kingston University, 2009.

Pascoe, Bruce. *Dark Emu: Black Seeds: Agriculture or Accident?.* Magabella Books, 2014.

Persig, Robert. *Zen and the Art of Motorcycle Maintenance.* Bantam, 1988.

Peterson, T. *Nightwork: A History of Hacks and Pranks at MIT* (updated edition). MIT Press (2001).

Petrich, Mike and Karen Wilkinson. *The Art of Tinkering: Meet 150+ Makers Working at the Intersection of Art and Science*. Weldon Owen, 2014.

Polakov, Claire. "Crafts and the Concept of Art in Africa", *African Arts*, vol. 12 no. 1 (November 1978): 22-23, 106.

Polanyi, Michael. *The Tacit Dimension*. Routledge, 1967.

Pollan, Michael "The Way We Live Now: Produce Politics", *The New York Times Magazine*, 14 January 2001.

Poulsen, Kevin. "'Banned' Xbox Hacking Book Selling Fast", *Business Week,* May 12, 2003.

Priest, Steven, editor. *Jean-Paul Sartre: Basic Writings*. Routledge, 2001.

Probyn, Elspeth. "Writing Shame", *The Affect Theory Reader*, edited by Melissa Gregg and Gregory J. Seigworth. Duke University Press, 2010: 71-91.

Procter, M. "Measuring Attitudes", *Researching Social Life*, edited by N, Gilbert. SAGE, 1995: 116-34.

Purdue, Derrick, Jörg Durrschmidt, Peter Jowers and Richard O'Doherty. "DIY Culture and Extended Milieux: LETS, Veggie Boxes and Festivals", *The Sociological Review,* vol. 45 no. 4 (1997): 645-67.

Putnam, Tim. "'Postmodern' Home Life", *At Home, An Anthropology of Domestic Space*, edited by Irene Cieraad. Syracuse University Press, 1999: 144-52.

Pye, David. *The Nature and Art of Workmanship*. Bethel: Cambium Press, 1968.

Quiggin, John. "Start with the Household", *Amateur Media: Social, Cultural and Legal Perspectives*, edited by Dan Hunter, Ramon Lobato, Megan Richardson and Julian Thomas. Routledge, 2014: 27-32.

Rand, Kelly. "Steampunk is the New Green', *Crafting a Green World*, 30 June 2008, http://craftingagreenworld.com/2008/06/30/steampunknewgreen/, accessed 16 March 2009.

Rassey, Lou. "Can the Maker Movement Remake America?", *Techonomy*, 17 September 2013, http://techonomy.com/conf/13-detroit/the-new-techonomy/can-the-maker-movement-re-make-america/.

Redington, Susan Bobby. *Fairy Tales Reimagined*. Jefferson: McFarland & Co., 2009.

Reserve Bank of Australia. "The Global Financial Crisis and Its Impact upon Australia", *Year Book Australia 2009-2010*. Canberra: Australian Bureau of Statistics, 2010.

Robertson, Kirsty. "Titanium Motherships of the New Economy: Museums, Neoliberalism and Resistance", *Imagining Resistance: Visual Culture and Activism in Canada*, edited by Kirsty Robertson and Keri Cronin. Wilfrid Laurier University Press, 2010: 207.

Robinson, Andrew and Simon Torney. "Utopias without Transcendence? Post-left Anarchy, Immediacy, and Utopian Energy", *Globalisation and Utopia: Critical Essays*, edited by P. Hayden, Palgrave Macmillan, 2009.

Rosen, Christine. "Technologies of the Self'", *The New Atlantis: A Journal of Technology & Society* (Summer 2005): 4-27.

Rosner, Jonathan and Daniela Bean. "Learning from IKEA Hacking: 'I'm Not One to

Decoupage a Tabletop and Call It a Day'", *CHI 2009 ~ Creative Thought and Self-Improvement*. Boston (2009).
Rundle, Guy. *A Revolution in the Making: 3D Printing, Robots and the Future*. Affirm Press, 2014.
Ruskin, John. "The Nature of Gothic", *Unto This Last and Other Writings*. Penguin Classics, 1997.
Ruskin, John. *Unto This Last and Other Writings*. London: Penguin, 1996.
Russell, Chris (presenter), "Going with the Flow", *Australian Story*, 26 October 2015, transcript available at http://www.abc.net.au/austory/content/2015/s4336886.htm.
Rutger, Bregman. *Utopia for Realists*. Bloomsbury 2017.
Rybczynski, Witold. *Home: A Short History of an Idea*. London: Pocket Books, 1986.
Sahlins, Marshall. *Culture and Practical Reason*. University of Chicago Press, 1976.
Sale, Kirkpatrick. *Rebels against the Future: The Luddites and Their War on the Industrial Revolution*. Perseus Publishing, 1995.
Sartre, Jean-Paul. *Being and Nothingness: The Principal Text of Modern Existentialism*. Translated Hazel E Barns. Washington Square Press, 1992.
Sayers, Jentery. "Tinker-centric Pedagogy in Literature and Language Classrooms", *Collaborative Approaches to the Digital in English Studies*, edited by Laura McGrath. Utah State University Press, 2011.
Schumacher, Ernst Friederich. *Good Work*. Cape, 1977.
Schwartz, David T. *Consuming Choices: Ethics in a Global Consumer Age*. Rowman & Littlefield, 2010.
Searle, Denise, editor. *Gathering Force: DIY Culture – Radical Action For Those Tired of Waiting*. London: The Big Issue Writers, 1997.
Seely-Brown, John. Igniting Innovation and Mastering Change conference, *Fora TV*, 16 November 2011. http://library.fora.tv/conference/igniting_innovation.
Seely-Brown, John. "The Power of Tinkering", *By Design*. ABC Radio National, 2 March 2011.
Seely-Brown, John. "Tinkering as a Mode of Knowledge Production in a Digital Age", Convening held at The Carnegie Foundation for the Advancement of Teaching, Stanford, California, 2008.
Segal, Stephen H. "Five Thoughts on the Popularity of Steampunk", *Weird Tales*, 17 August 2008, http://weirdtales.net/wordpress/2008/09/17/five-thoughts-on-the-popularity-of-steampunk/.
Sennett, Richard. *The Craftsman*. New Haven: Yale University Press, 2008.
Sennett, Richard. "Humanism", *The Hedgehog Review*, vol. 13 no. 2 (Summer 2011).
Seuss, Dr. *Sneetches on Beaches*, Random House, 1989.
Shiner, Larry. *The Invention of Art: A Cultural History*. The University of Chicago Press, 2001.
Shirky, Clay. *Cognitive Surplus: Creativity and Generosity in a Connected Age*. New York: Penguin, 2008.
Shove, Elizabeth. *Comfort, Cleanliness + Convenience: The Social Organisation of Normality*. Oxford: Berg, 2003.

Shove, Elizabeth, Frank Trentmann and Richard Wilk, editors. *Time, Consumption and Everyday Life: Practice, Materiality and Culture*. Oxford: Berg, 2009.

Shove, Elizabeth, Matthew Watson, Martin Hand and Jack Ingram, editors. *The Design of Everyday Life*. Oxford: Berg, 2007.

Silver, Charles and Nathan Jencks. *Adhocism: The Case for Improvisation*. New York: Anchor Books, 1973.

Silver, Steven and Randal Verbrugge. "Home Production and Endogenous Economic Growth", *Journal of Economic Behavior & Organization*, vol. 75 no. 2 (August 2010).

Slater, Don. *Consumer Culture and Modernity*. Cambridge: Polity, 1997.

Slater, Don. "The Ethics of Routine", *The Design of Everyday Life*, edited by Elizabeth Shove, Matthew Watson, Martin Hand and Jack Ingram. Oxford: Berg, 2007: 217-30.

Smith, Terry. *Thinking Contemporary Curating*. Independent Curators International, 2012.

Snowdon, Warren. "$10,000 up for Grabs for Men's Sheds", Press Release, M.F.I. Affairs, ed. House of Representatives, Canberra (2011).

Söderberg, Johan. "Wealth Without Money, One 3D Printer at a Time: The Cunning of Instrumental Reason", *Journal of Peer Production*, no. 4 (January 2014).

Spencer, Amy. *DIY: The Rise of Lo-Fi Culture*. London: Marion Boyars, 2005.

Spencer, Nicholas. "Rethinking Ambivalence: Technopolitics and the Luddites in William Gibson and Bruce Sterling's *The Difference Engine*", *Contemporary Literature*, 22 September 1999.

Springwood, Charles F. and C. Richard King. "Unsettling Engagements: On the Ends of Rapport in Critical Ethnography", *Qualitative Research Methods*, edited by Paul Atkinson and Sara Delamont. SAGE, 2010: 165-72.

Starn, Orin. "Writing Culture at 25: Special Editor's Introduction", *Cultural Anthropology*, vol. 27 no. 3 (August 2012): 411-16.

Stebbins, Robert A. *Amateurs: On the Margin between Work and Leisure*. SAGE, 1979.

Stebbins, Robert A. "Educating for Serious Leisure: Leisure Education in Theory and Practice", *World Leisure and Recreation*, vol. 41 no. 4 (1999): 14-9.

Stebbins, Robert A. "Right Leisure: Serious, Casual, or Project-based?", *NeuroRehabilitation*, vol. 23 no. 4 (2008): 335-41. (2008B).

Stebbins, Robert A. *Serious Leisure: A Perspective for our Time*. New Brunswick: Transaction Publishers, 2008 (2008A).

Steel, Sharon. "Steam Dream", *The Boston Phoenix*, 19 May 2008, accessed 13 July 2009, http://thephoenix.com/boston/Life/61571-Steam-dream/?page=2.

Steil, Michael. "17 Mistakes Microsoft Made in the Xbox Security System", paper delivered at the 22nd Chaos Communication Congress, December 29, 2005, http://www.xbox-linux.org/wiki/17_Mistakes_Microsoft_Made_in_the_Xbox_Security_System.

Stephens, Tony. "A Little Bit of Tinkering, and Old Wagon Heads For a Home", *Sydney Morning Herald*, 20 July 2002.

Sterne, Jonathan. "Communication as Techne", *Communication as… Perspectives on Theory*, edited by Gregory J. Shepherd and Jeffrey St. John. SAGE, 2006.

Stewart, Susan. *On Longing: Narratives of the Miniature, the Gigantic, the Souvenir, the Collection*. Duke University Press, 1984.

Stone, Wendy. "Downshifter Families' Housing and Homes: An Exploration of Lifestyle Choice and Housing Experience", PhD, Swinburne University of Technology, 2010.

Strathern, Marilyn. "Externalities in Comparative Guise", *The Technological Economy*, edited by Barry et al. Routledge, 2005: 66-82.

Stretton, Hugh. *Economics: A New Introduction*. Pluto Press, 1999.

Summers, Anne. *The End of Equality: Work, Babies and Women's Choices in 21st Century Australia*. London: Random House, 2004.

Taddeo, Julie Anne and Cynthia J. Miller. *Steaming Into a Victorian Future: A Steampunk Anthology*. Scarecrow Press, 2013.

Taleb, Nassim Nicholas. *Antifragile: Things that Gain from Disorder*. Random House, 2012.

Thomas, Douglas and John Seely-Brown. *A New Culture of Learning: Cultivating the Imagination for a World of Constant Change*. Soulellis Studio, 2011.

Thomson, Mark. *Blokes and Sheds: Stories from the Shed* (combined edition). Sydney: Angus & Robertson/HarperCollins, 2002. (2002A).

Thomson, Mark. "I Tinker Therefore I Am", *The Haystack Reader: Collected Essays on Craft 1991-2009*, edited by M. Alpert. Maine: University of Maine Press/ Haystack Mountain School of Crafts, 2010: 347-56.

Thomson, Mark. *Makers, Breakers and Fixers: Inside Australia's Most Resourceful Sheds*. Sydney: HarperCollins, 2007.

Thomson, Mark. *Rare Trades: Making Things by Hand in a Digital Age*. Sydney: HarperCollins, 2002. (2002B).

Thomson, Mark. 'Tinkering' (blog post), *The Website of The Institute of Backyard Studies: Home of Shed Culture*, 2009, http://www.ibys.org/shed/?p=543, accessed 1 May 2011.

Thrift, Nigel. "*Fings Ain't Wot They Used t' Be*: Thinking through Material Thinking as Placing and Arrangement"', *The Oxford Handbook of Material Culture Studies*, edited by Mary C. Beadry and Dan Hicks. Oxford University Press, 2010: 634-45.

Throsby, David and Virginia Hollister. *Don't Give Up Your Day Job: An Economic Study of Professional Artists in Australia*. Australia Council for the Arts (2003).

Tickel, Joshua, Kaia Tickel and Kaia Roman. *From the Fryer to the Fuel Tank: The Complete Guide to Using Vegetable Oil As An Alternative Fuel.* Tickle Energy Consulting, 2000.

Tilley, Chris, Webb Keane, Susanne Küchler, Mike Rowlands and Patricia Spyer, editors. *Handbook of Material Culture*. SAGE, 2006.

Timms, Peter. *What's Wrong with Contemporary Art?*. Sydney: University of New South Wales Press, 2005.

Tofts, Darren. *Paradise Lost or Utopia Regained?*, keynote presentation from the Melbourne launch of the Experimenta International Biennial of Media Art: *Experimenta Utopia Now*, Melbourne, Australia, February 2010.

Travis, Joseph Meinolf. "Common Goods", *Utopian Studies* (2011): 352-7.
Trimble, Tyghe. "Why Tinkering Is Always Time Well Spent", *Popular Mechanics*, 24 May 2010.
Tulley, Gever. "Gever Tulley Teaches Life Lessons through Tinkering", TED talks, 2009.
Van Ekeren, Glenn. *Tinker: The Art of Challenging the Status Quo*. Higherlife Development Services, 2014.
Vandermeer, Ann and Jeff Vandermeer, editors. *Steampunk*, Tachyon, 2008.
von Busch, Otto. "Molecular Management: Protocols in the Maker Culture", *Creative Industries Journal*, vol. 5 no. 1-2 (2012).
Von Slatt, Jake, *The Steampunk Workshop*, at http://steampunkworkshop.com/.
Waksman, Steve. "California Noise: Tinkering with Hardcore and Heavy Metal in Southern California", *Social Studies of Science*, vol. 34 no. 5 (2008): 675-702.
Wallace, Jacqueline. "Yarn Bombing, Knit Graffiti, and Underground Brigades: A Study of Craftivism and Mobility", *Journal of Mobile Media*, vol. 6 no. 3 (2013).
Walter, Damien G. "Steampunk: The Future of the Past", *The Guardian* (UK edition), 7 January 2009.

Ward, Russel. *The Australian Legend*. Oxford University Press, 1965.
Wark, McKenzie. *A Hacker Manifesto*. Cambridge: Harvard University Press, 2004.
Wark, McKenzie. "Hackers", *Theory, Culture, Society*, vol. 23 no 2-3 (March-May 2006): 320-2.
Wark, McKenzie. "A More Lovingly Made World", *Cultural Studies Review*, vol. 19 no. 1 (March 2013): 296-304.
Warner, Michael. *Publics and Counterpublics*. Zone Books, 2005.
Waterhouse, Richard. *The Vision Splendid: A Social and Cultural History of Rural Australia*. Curtain University Books, 2005.
Waterton, Claire. "Amateurs as Experts: Re-thinking Expertise with Non-Expert Knowledges", *ESRC Research Report*, 2006, http://www.esrcsocietytoday.ac.uk.
Watkins, Evan. "Social Position and the Art of Automobile Maintenance", *Throw-aways: Work Culture and Consumer Education*. Stanford University Press, 1993: 89-96.
Watson, Don. *The Bush: Travels in the Heart of Australia*. Penguin Books, 2014.
Watson, Matt and Elizabeth Shove. "Product, Competence, Project and Practice: DIY and the Dynamics of Craft", *Journal of Consumer Culture*, vol. 8 no. 1 (2008): 69-89.
Wegner Etienne and Richard McDermott. *Cultivating Communities of Practice*. Harvard Business School Publishing, 2002.
Weidel, Janine. *Irish Tinkers*. Palgrave Macmillan, 1979.
Wilkinson, Karen and Mike Petrich. *The Art of Tinkering*. Weldon Owen, 2014.
Williams Colin, Andrew Leyshon and Roger Lee. *Alternative Economic Spaces*. SAGE, 2003.
Williams, Colin C. *The Hidden Enterprise Culture*. Edward Elgar Publishing, 2006.
Williams, Colin C. "A Lifestyle Choice? Evaluating the Motives of Do-It-Yourself

(DIY) Consumers", *International Journal of Retail Distribution and Management*, vol. 32 no. 4 (2004): 270-9. (2004A).
Williams, Colin C. "The Myth of Marketization: An Evaluation of the Persistence of Non-Market Activities in Advanced Economies", *International Sociology*, vol. 19 no. 4 (2004): 437-9. (2004B).
Williams, Colin C et al (John Round, Peter Rodgers). "Beyond the Formal/Informal Economy Binary Hierarchy", *International Journal of Social Economics*, vol. 34 no. 6 (2007): 402-14.
Williams, Kristen A. "Old Time Mem'ry: Contemporary Urban Craftivism and the Politics of Doing-It-Yourself in Postindustrial America", *Utopian Studies,* vol. 22 no. 1 (2011): 303-20.
Wilson, Betty. "'Make Do and Mend' in Bomb-torn London", *Sydney Morning Herald*, 21 February 1945.
Wilson, Frank R. *The Hand: How Its Use Shapes the Brain, Language and Human Culture*. Vintage, 1998.
Wilson, Katherine. "A New Steam of Consciousness", *The Age*, 22 May 2010. (2010A).
Wilson, Katherine. "The Rhythm of Engagement," *Overland*, no. 201 (2010): 14-21. (2010B).
Wilson, Katherine. "Steampunk," *Meanjin*, vol. 69 no. 2 (2010): 20-33. (2010C).
Wilson, Katherine. "When 'hand-crafted' is really just crafty marketing", *The Conversation*, 21 September 2015.
Winner, Langdon. "Do Artifacts Have Politics?" *Daedalus*, vol. 109, no. 1 (Winter, 1980): 121-36.
Wodak, Ruth, and Michael Meyer. *Methods of Critical Discourse Analysis*. SAGE Publications, 2001.
Woolgar, Steve. "Technologies as Cultural Artefacts", *Information and Communication Technologies: Visions and Realities*, edited by W. H, Dutton and M, Peltu. Oxford University Press, 1996: 87-102.
Yuval-Davis, Nira, Kalpana Kannabiran, and Ulrike Vieten, editors. *The Situated Politics of Belonging*. SAGE, 2006.
Zeldin, Theodore. *An Intimate History of Humanity*. Vintage, 1994.

About the Author

Katherine Wilson works as an editor, writer, researcher and tinkerer. She also teaches in the university sector, where she has a PhD in cultural studies. Her feature articles have appeared in the *Age*, *Sydney Morning Herald*, *Conversation*, *Australian*, *Courier-Mail*, *Art Monthly*, *Crikey.com*, *New Matilda*, the *Law Institute Journal* and *Good Weekend*. Her essays have appeared in journals including *Griffith Review*, *Meanjin*, *Eyeline*, *Eureka Street* and *Overland*. She edited *Overland* between 2002 and 2007 and has worked in advocacy roles for non-profit and environmental bodies.